TOUCHING SPACE, PLACING TOUCH

Touching Space, Placing Touch

Edited by

MARK PATERSON
University of Pittsburgh, USA

MARTIN DODGE
University of Manchester, UK

Routledge
Taylor & Francis Group

LONDON AND NEW YORK

First published 2012 by Ashgate Publishing

Published 2016 by Routledge
2 Park Square, Milton Park, Abingdon, Oxfordshire OX14 4RN
711 Third Avenue, New York, NY 10017, USA

First issued in paperback 2016

Routledge is an imprint of the Taylor & Francis Group, an informa business

British Library Cataloguing in Publication Data
Touching space, placing touch.
1. Touch. 2. Touch–Therapeutic use. 3. Human geography. I. Paterson, Mark, 1972– II. Dodge, Martin, 1971–
152.1'82–dc23

Library of Congress Cataloging-in-Publication Data
Paterson, Mark, 1972–
Touching space, placing touch / by Mark Paterson and Martin Dodge.
   p. cm.
Includes bibliographical references and index.
ISBN 978-1-4094-0214-5 (hardback : alk. paper)
1. Touch—Psychological aspects. 2. Touch—Social aspects. 3. Spatial behavior. 4. Human behavior. 5. Human geography. I. Dodge, Martin, 1971– II. Title.
BF275.P39 2012
152.1'82—dc23

2012012178

ISBN 13: 978-1-138-25349-0 (pbk)
ISBN 13: 978-1-4094-0214-5 (hbk)

Contents

List of Figures and Tables

Figures

Tables

Acknowledgments

Touching Space, Placing Touch derived originally from sessions at the RGS-IBG annual conference for geographers in London in 2007, organised by Sara MacKian, Martin Dodge, Chris Perkins, and Mark Paterson. The sessions seemed to ride on the crest of a wave of interest not solely concerned with the culture and spatiality of the sense of touch *per se*, but also with the social, therapeutic and gendered aspects of tactile encounters and sensual interactions. The conference sessions revealed a healthy diversity of disciplinary and methodological approaches and clearly demonstrated that touch, in its many performances, articulations and practices, was a vibrant area of ongoing research. This was reflected not only by the quality of papers presented but also by the appreciative and engaged audience who supplied numerous questions and critical suggestions. Of course, great conference sessions do not automatically make good books. Since the conference, this edited collection has experienced a somewhat sinuous and protracted development while under the generous stewardship of Ashgate. There have been some inevitable modifications, subtractions and additions to what was presented at the 2007 meeting, but such changes readily indicate that interest in touch within the social sciences and humanities has not abated.

The editors would now like to express their thanks. To Valerie Rose, Senior Commissioning Editor at Ashgate, for her continued support of this book project. To all the contributors for their forbearance in the somewhat protracted editing and production of this volume. Some were involved right from the start and contributed in a very timely manner, while some were involved at a much later stage, under pressure of deadlines. To the former, we apologise for the delay; to the latter, we are grateful that you came on board and made this collection what it is now. But most of all, to Sara MacKian, who was instrumental in initiating the conference sessions and the early stages of the book proposal. We are extremely grateful to her for continuing to be involved, looking over the contributors' texts, writing her own chapter, and co-writing the Introduction. Sara's presence has therefore been integral to the total life of this project and would not have happened without her.

We are grateful to Taylor and Francis. Some of the material for Rachel Colls' chapter is developed from her article 'BodiesTouchingBodies: Jenny Saville's over-life-sized paintings and the 'morpho-logics' of fat, female bodies', *Gender Place and Culture* Vol. 19, No. 2: 175–192.

Notes on Contributors

Bernard Andrieu

Bernard is a philosopher of the body. He edited *The Dictionary of the Body* (CNRS 2006), co-founded and edits the journal *International Review of the Body*, and has developed a wide range of work about body ecology, touch and hybridity for the description of the lived body, and the ecological constitution of the self through interaction with others. His key writing includes 'Technophobie ou ecotechno' (*Australian Journal of French Studies*, 2011), 'Trans-body and hybrid identity' (*Scan: Journal of Media Arts Culture*, 2007), the chapter 'Embodying the chimera: Toward a phenobiological subjectivity' in *Signs of Life, BioArt And Beyond* (MIT Press 2007), and 'Brains in the flesh: Prospects for a neurophenomenology' in *Janus Head: Journal of Interdisciplinary Studies in Literature, Continental Philosophy, Phenomenology, Psychology and Arts* (2006). His blog is at http://leblogducorps.over-blog.com.

Amanda Bingley

With a background in psychotherapy and psychoanalytic geographies, Amanda's research interests lie in the relationship between mental health, well-being and place, and includes work with older people and the benefits of gardening, the effects of woodland spaces on mental health in young people, and various projects in palliative care. She has published key work in *Social and Cultural Geography*, *Children's Geographies*, *Social Science & Medicine*, *Palliative Medicine* and contributed to several edited volumes including *Emotional Geographies* (Ashgate 2005) and *Making Sense of Place: Multidisciplinary Perspectives* (Boydell Press 2012).

Sarah Cant

Sarah is a cultural geographer interested in cultures of creativity, improvisation and dance, with a particular focus on ethnography and observant participation. Her research is informed by interests in embodiment, sensuous experiences/knowledges, and feminist interpretations of the body and relationships. Projects have included work on public art, improvised performance and theatre, and she is currently researching social dance, specifically Argentine tango in the context of British dance communities.

Rachel Colls

Rachel's research is concerned with theorising and researching 'the body', and her work is largely informed by engagements with feminist post-structuralist theory including that of Elisabeth Grosz, Judith Butler and Luce Irigaray. She has published on topics including non-representational geographies, women's embodied experiences of clothing consumption and children's healthy food practices. Her most recent work has focused on developing critical geographies of obesity and has involved research concerned with the practices of weighing and measuring children, size accepting spaces and fat materialities.

Martin Dodge

Martin's research focuses on conceptualising the socio-spatial power of digital technologies and virtual geographies, and the theorisation of visual representations, cartographic knowledges and novel methods of geographic visualization. He curated the well-known web-based Atlas of Cyberspaces and has co-authored three books covering aspects of the spatiality of computer technology: *Mapping Cyberspace* (Routledge 2000), *Atlas of Cyberspace* (Addison-Wesley 2001) and *Code/Space* (MIT Press 2011). He's also co-edited three books, *Geographic Visualization* (Wiley 2008), *Rethinking Maps* (Routledge 2009) and *The Map Reader* (Wiley-Blackwell 2011), all focused on examining the social and cultural meanings of new kinds of mapping practice.

Rob Kitchin

Rob is Professor of Human Geography and Director of the National Institute of Regional and Spatial Analysis (NIRSA) at the National University of Ireland, Maynooth and Chair of the Management Board of the Irish Social Sciences Platform (ISSP). He has published twenty books, is editor of the international journal *Progress in Human Geography*, and co-editor-in-chief of the *International Encyclopedia of Human Geography* (Elsevier 2009).

Alexandre Klein

Alexandre is a philosopher and historian of sciences. After two years of teaching in health education he is currently finishing a doctoral thesis about the subject's epistemology in French contemporary medicine. His research focuses on relations between body representations and identity building, particularly in health practices. He has published work on the history, philosophy, epistemology and ethics of medicine, and has recently edited a volume about health sensations, *Les Sensations de Santé* (Nancy University Press 2010).

Anne-Flore Laloë

Anne-Flore's research is concerned with the historical geographies of oceanography and marine sciences. She is currently pursuing these at the Marine Biological Association of the UK. She is also an honorary fellow at the University of Exeter. In a previous position, however, she worked as a French-language medical interpreter and translator for the National Health Service. It is in this capacity that she contributed to the chapter *Touch, Skin Cultures and the Space of Medicine* in this volume, co-authored with Bernard Andrieu and Alexandre Klein.

Jennifer Lea

Jennifer is interested in geographies of the body and embodiment. She is currently working on an ESRC-funded project looking at the social relationships of children with special educational needs in mainstream and special schools. She is also involved in an AHRC-funded project which looks at everyday spiritual practice such as yoga and meditation. Her previous research explored health, care and embodied learning through the practices of therapeutic massage and yoga. Her work has been published in international journals such as *Body & Society* and *Geoforum*.

Jamie Lorimer

Jamie's research develops interdisciplinary approaches to wildlife conservation. These need not appeal to problematic understandings of a pure Nature, removed from Society and revealed through objective Natural Science. Instead his work combines social and natural sciences to offer a new approach to biogeography that is sensitive to the hybrid character of the environment and the political nature of claims for environmental knowledge. Drawing on fieldwork in the UK, Europe and South Asia he presents ways of theorising and practicing human-environment relations better adapted to the dynamic, diverse and cosmopolitan ecologies of the twenty-first century.

Sara MacKian

Senior Lecturer in Health and Wellbeing at The Open University, and formerly Lecturer in Health Geographies at the University of Manchester, the driving theme underpinning Sara's research is a curiosity for how people, communities and organisations interact around issues of illness, health and well-being. This has led to a range of studies exploring ME, parenting, sexuality, spirituality and aspects of public health. A geographer by training, she has a particular interest in qualitative research methodologies. She is author of *Everyday Spirituality: Social and Spatial Worlds of Enchantment* (Palgrave Macmillan 2012).

Hannah Macpherson

Hannah's research draws on feminist, post-structural and post-phenomenological theories and contributes to debates on embodied ethics, disability, visual impairment, concepts of landscape and the methodological challenges of representing research encounters. She has numerous peer-reviewed publications including work in *Environment and Planning A*, *Society and Space*, *Cultural Geographies* and *The Senses & Society*. She is co-ordinator of the Disability Ethics and Aesthetics research group at the University of Brighton and is currently conducting research with performance artists with disabilities.

Pau Obrador

Pau has co-edited *Cultures of Mass Tourism* (Ashgate 2009) and has published work in a range of tourism and cultural geography journals. His research interests are in the theorisation of tourism, cultures of mass Mediterranean tourism and the corporeality of the tourist experience. His work on nudity and the beach explores the role of the senses in culture and society, with a special emphasis on touch. He is currently working on various aspects related to mass tourism, family tourism and tourism futures.

Mark Paterson

Based at the University of Pittsburgh, Mark's research centres around haptics, the role of the somatic senses, and extensions of the sensorium through technology. His books include *Consumption and Everyday Life* (Routledge 2005), *The Senses of Touch: Haptics, Affects and Technologies* (Berg 2007), and is currently finishing a book entitled *Seeing with the Hands: A Philosophical History of Blindness* for Reaktion. He has published work in a range of humanities and social science journals and chapters in books on touch, haptics and blindness, and has undertaken funded research projects in the areas of robot skin and the haptic modelling of prehistoric textiles. His research website is http://www.sensesoftouch.co.uk.

Elizabeth Straughan

Elizabeth graduated with a Bachelors Degree from the University of Wales, Aberystwyth and went on to do a Masters and a Ph.D. at the same institute. Focusing on the body through an attendance to the skin as well as the material and metaphorical dynamics of touch, her research seeks to unravel the volatile nature of the body enhanced, manipulated and transformed by technologies, Elizabeth is currently working as an AHRC/NSF funded post-doctoral research fellow on a project exploring art-science collaborations at Aberystwyth University's IGES department.

Anne Volvey

Anne's research deals with contemporary art and with the epistemology of geography. Fieldwork is the common concern in her research, looking at fieldwork practices and experiences as transitional processes, and elaborating upon these processes through the use of English and French transitional psychoanalytic theories. She has organized international conferences and has (co-)edited books and journal special issues in these areas. They include *Activité artistique et spatialité* (L'Harmattan 2010) and *Terrains de Je. (Du) Sujet (au) géographique* (*Annales de Géographie,* 5(687), 2012).

Introduction: Placing Touch within Social Theory and Empirical Study

Mark Paterson, Martin Dodge and Sara MacKian

> Like the air we breathe, [touch] has been taken for granted as a fundamental fact of life, a medium for the production of meaningful acts, rather than meaningful in itself. (Classen 2005, 2)

Placing the Senses

So, where has touch been within social theory and spatial scholarship all this time? Where is it now and where might it be placed in the future? What kinds of knowledges are produced, validated and employed in researching the spaces of touch and the places of touching in different social contexts? These are the central concerns of *Touching Space, Placing Touch*. Before focussing on those distinctive aspects of touch, let us first consider the troubled place of the senses in general.

'The [origin of all thoughts] is that which we sense, for there is no conception in Man's [*sic*] mind which hath not at first totally or in parts, been begotten upon the organs of sense', declares Thomas Hobbes in the opening chapter of his celebrated 1651 work of political philosophy *Leviathan* (1962, 21). His formulation evokes a deep, pervasive channel running throughout Western philosophy, from pre-Socratic thinking, through Enlightenment debates around rationalism and empiricism, right up to contemporary Poststructuralist concerns, in considering grounds for the relationship between sensory experiences and the formulation of more complex knowledge and ideas. Hobbes' reasoning involves building from first principles, beginning by understanding the most immediate and seemingly straightforward components of individual experience, sensation, and building from there into more complex social ideas. Hobbes claimed we were subject to two types of phenomena, in his parlance 'sensation' and 'imagination' (thought). From this, Hobbes formulates a more realistic hypothesis of a 'commonwealth' as the principle of a just social order constituted by an artificial collective of people. Starting from those most basic units of human experience, sensations and thoughts, a political philosophy of a fairer society is formed, connecting the individual sensorium to a larger social order. Any accusation that sensory knowledge is trivial, ephemeral or 'merely' subjective therefore misses the point.

John Locke (1690) broadly agreed with Hobbes' thesis, claiming that the entirety of human experience was derived from two sources, sensation and reflections. 'This great Source, of most of the Ideas we have, depending wholly on our Senses, and derived by them to the Understanding, I call Sensation' (Locke 1975, 105). The assumption that the senses were the foundation of individual experience, an epistemology of corporeal objects rather than spiritual ideas, is a tenet of empiricism as opposed to the earlier rationalism of Descartes and others, but also consistent with earlier medieval philosophy. The broad and pervasive consensus throughout Western philosophy from Plato onwards is also exemplified by one of the founding fathers of Christianity, Saint Augustine, who also starts from 'first knowledge' of corporeal beings through the senses in order to attain higher knowledge of spiritual matters, more permanent knowledge 'towards God' (1950, 109). Such hierarchies of knowledge, with the preliminary nature of the senses, is clearly well established, despite the acknowledgement that 'first knowledge' of the body and senses is unreliable, susceptible to biases, and only a starting point on the much longer journey to 'higher knowledge' and apotheosis. This epistemological model is unmistakable and pervasive, illustrated and instantiated within literary tropes and high art alike. For example the famous Renaissance painting *The Allegory of Touch* (Figure I.1), which Harvey (2011, 393) declares represents 'the nexus between the body and its affective life (being emotionally "touched"), the medical and anatomical understanding of skin as both a bodily covering and a receptor of touch... and the mythological narratives about touch that undergird early modern culture'. Further illustrations include Peter Damian in the eleventh century, likening each of the senses to 'five vulnerable and poorly guarded gates of a city' (in Jütte 2005, 77)

The purpose of this historical synopsis is not to construct an all-encompassing argument or assert any grand narrative for *the* place of the senses in Western history, philosophy or cultural life. Instead, we wish to make three substantive points. Firstly, to foreground and contest the historical pervasiveness of a model of thinking that tries to bypass the importance of immanent sensory knowledges in order to assert the superiority of 'higher' wisdom, or transcendent truths. Secondly, to contest any unitary or easily universalistic conception of 'place', 'sense' or 'touch' that such a template might assume, signalled by the diversity of topics and approaches within this edited collection. The contributors to *Touching Space, Placing Touch* derive their approaches from cultural geography, art history, psychotherapy, social theory, empirical fieldwork-based social science, and much else besides. The multiplicity of ways that 'sense', 'touch' and the diversity of 'places' wherein these are encountered, belies any such generalising assumption. Thirdly, and most significantly, we make the case that the primacy and living immediacy of sensory experience does not reside solely within the boundaries of the skin, somehow locked within discrete, disconnected bodies. This is why the historic narrative of Hobbes' *Leviathan* is worth reprising. The *senses* are not equivalent to the tissues and cells of the sense organs themselves, nor reduced to nerves that connect to the brain. The cultural chronology of the formulation of a 'sensorium' necessitates that the senses are ineluctably social: felt individually, but also *always* shared intersubjectively. A sensorium is the sum of an organism's perception, the seat of sensation, and 'the subject's way of coordinating all the body's perceptual and proprioceptive signals as well as the changing sensory envelope of the self' (Jones 2007, 8). Although physiologically

Figure I.1 *The Allegory of Touch,* **painted by Jan Brueghel the Elder and Peter Paul Rubens (Ca. 1617; oil on panel 65cm x 110 cm)**

Source: courtesy of Museo Nacional del Prado, Madrid

located within an individual body, its operation is continually shifting and culturally variable. As Hobbes and others have detailed, complex knowledges originate from the position that the sense modalities are a necessary prerequisite for experiences of embodied consciousness and are the principal source of contact with the world for corporeal beings. By implication, then, there are congruent slippages in that consciousness as a result of sensory impairments of any kind (say from processes of aging, illness or genetic inheritance), and humans are always open to experiencing the social and spatial world differently as a result.

Yet often there remains a tendency to take the senses as given, or somehow superfluous or inconsequential, especially in social science scholarship. The senses are relegated to common sense or parcelled off as automatic biological function, unworthy of more detailed social exploration and nuanced explanation. While so central to the embodied experience of researcher and researched alike, there has been surprisingly little reflexivity about the role of the senses in the actual practice of *doing* research in the social sciences. As we shall see, in a post-embodiment scholarship alert to 'more-than representational' sensibilities in recent years, a renewed interest and concern with the senses, including touch, is beginning, part of a collective upsurge of research across several disciplines.

An increased attention to touch and its modalities necessarily results in a widening array of attendant research questions. For example, should touch be researched as a unitary sense or modality? Is touch straightforwardly cutaneous, a surface feeling upon the skin, and how far might it be related to other, less distinct, sensations within the body? 'Things are quite simple until a scientist comes along and complicates them', the biologist Otto Lowenstein (1966, 121) once declared, having conducted pioneering laboratory experiments on the pressure sensors in different animals in the 1950s and 1960s. The way that the senses of balance, movement and bodily orientation in space were constituted through sensing cells and organs distributed throughout the body of animals and humans alike, owes much to his research. Another way of characterising such research was, as Lowenstein (1966, 121) himself pithily put it, in converting 'common sense' to 'uncommon sense', yet with far-reaching and unforeseen implications. Within experimental biology and the near-contemporary field of Gibson's (e.g. 1968) ecological psychology, the role of detailed laboratory findings in challenging and contesting previously straightforward and long held philosophical assumptions about the sensory modalities, their neurological pathways and information channels was crucial. In a similar vein, an agenda for rethinking divisions between sensory modalities and for grasping how they have been historically, culturally and socially formed is increasingly a concern for current social science. We hope that *Touching Space, Placing Touch* contributes in its own way to this grander project, turning common sense into uncommon sense by questioning assumptions about the senses, their felt experience as immediate and/or unmediated, their interaction, their role in the perception of space, and the role of the social in the formation of a sensorium. As with Lowenstein's characterisation, such assumptions had been largely unchallenged until recent social scientific scholarship came along to try to complicate them.

The Place of Touch and a Renewed Interest the Body

With the 'turn to the body' in social theory in previous decades and the so-called 'cultural turn' in human geography and anthropology, some claims have been made about a 'return to the senses' (cf. Paterson 2008), marked by the rise of a transdisciplinary field known as 'sensory studies' that connects developments across a number of academic areas and methodological approaches, weaving historical, theoretical, and empirical study into something rich, relevant and potentially revealing (see http://sensorystudies.org for an ongoing database of scholars compiled by anthropologist David Howes' team). If the senses have previously been largely the preserve of biology and experimental psychology, a 'sensory turn' across a number of fields in the humanities and social sciences can be identified which involves the examination or re-examination of the senses according to the conceptual specificities and methodological limitations of each discipline. Not an intellectual movement as such, it is more a collation of a series of parallel strands threaded through and interconnecting with much larger disciplinary histories, and we will examine the specific story of human geography in this regard below. The development of book series and new journals, including *Senses & Society* (founded by David Howes in 2005), speak to the transdisciplinary potential of taking the senses seriously in the humanities and social sciences.

Yet, while there is evident promise within a transdisciplinary 'sensory studies' the scope for original findings is, we believe, more questionable. This is because the weave between the theoretical complexity necessary to do justice to human senses, sensibilities and bodily dispositions is not often matched with the kinds of flexible and nuanced empirical approaches required to do them justice. Moreover, the role of the senses within academic research has not always been explicitly demarcated, nor the sole focus of study. Sensory experience has obviously been implicit, pervasive within research activities and therefore inherently present in some form or another, but rarely recorded or deemed worthy of analysis in itself. Linked to this trend, if the late twentieth century interest in embodiment was characterised by thinking about the body as a site of signification for the politics of gender, or the production of meaning through adornment, inscription and so on, the early twenty-first century is seeing another blooming of interest in 'the body', this time as an explicit research tool. For, despite the vast quantity written about the body across academic disciplines in the 1980s and 1990s, very rarely was the body used intimately and reflexively as an actual instrument for 'doing' the research, as not simply the focus but the means through which social science investigation were conducted, something that geographers Crang (2003) and Longhurst et al. (2008 and 2009) have made a powerful case for.

Employing a cross-disciplinary approach in a post-embodiment context we can therefore identify a burgeoning area of work that addresses individual sensory experiences, yet which also remains conscious of the embeddedness of the senses in society, and in the spirit of Hobbes, how a sensorium is historically formed and socially co-constituted. Examples of some of the most compelling work include Lisa Law's (2001) article 'Home Cooking', whose ethnographic work amongst immigrant Filipino workers in Hong Kong evoked the smells and tastes of their cooking as an integral component in reproducing 'home'. Yet it must still be broadly acknowledged that there is much to do, within specific disciplines, to better attend to such concerns and address a sensory lacuna in their historicity.

Studies of visual culture and historical accounts of vision and socio-technical means of seeing have proliferated, most notably Crary (1990 and 1999), Mitchell (2002) and Danius (2002), demonstrating that the scopic has enjoyed a disproportionate amount of research interest compared to the other sense modalities. By comparison touch remains under-explored, under-represented and marginal across these broad categories of research. Nevertheless, an increasing amount of work is contesting this prioritisation of the visual, what Martin Jay in *Downcast Eyes* (1994) famously called 'ocularcentrism', or reconsiders the visual in relation to other modalities through renewed approaches to art history, film studies, literary studies, or traditional aesthetics for example. Re-examining the visual's relationship with the non-visual has been a concern within art history, architecture and cultural geography (e.g. Harvey's 2003, 2011 work on touch in art history; Pallasmaa's architectural theory, 2005; the "more-than visual" approach to the built environment of Paterson, 2010), thereby contributing perspectives that enfold visual *and* non-visual cultures. In anthropology the move to consider the embodied nature of fieldwork has already brought valuable attention to the non-visual modalities, such as the ethnographic monographs of Stoller (1997), Howes (2003) and Geurts (2002). While in sports science, Sparkes (2009) and Hockey (2006) for example are similarly connecting anthropology and psychology literature and attending to the more explicitly somatic processes involved in physical activity. By examining their disciplinary histories in parallel ways, establishing a corrective to a previously visual bias, or attempting to reconsider the relationship between the visual and the non-visual, a considerable amount of cross-pollination is taking place across academic fields which, in some cases, is reinvigorating existing debates. For example, in film studies the idea of haptic cinema has a hold, thanks in part to phenomenologically-inflected contributions from Sobchack (2004) and Marks (2002). Where technologies are involved, such as medicine or computer mediated communications, a shift in those relations may be considerable.

The Places of Touch in Geographical Scholarship

Where, then, is the place of touch? Until recently, social science research that dealt specifically with touch and tactility was thin on the ground. Moreover, given the deep importance of touch in all aspects of spatial experience, the tactile senses have been surprisingly poorly researched by human geography. It is evident that geographers have quite simply and literally been out of touch. There are many reasons why touch is an overlooked spatial practice. As suggested above, this neglect by geographers is part and parcel of orientations to the senses in general. The nature of touch is classified as immediate, obvious or trivial, yet it is hard to encode these intimate sensations and their subtle meanings into representational forms that prioritise text and the print medium, the usual means of outputting academic research. Furthermore it is an under-theorised sense in geography. Perhaps, as Rose (2003) has argued, the heart of the geographic enterprise is historically visual, originating from scientific cultures of detachment and observation during fieldwork. Furthermore, the processes of data collection and manipulation involved in GIS (Geographic Information System) similarly works by abstracting data from the inevitably embodied processes of collecting and collating, representing it in primarily visual terms through digital cartography.

Given the return to the body in previous decades, the intermittent scholarly flirtations with phenomenology in sociology, anthropology and human geography in the 1980s, and the more recent upsurge of interest in the body as an instrument of research, whether through feminist programmes of research (e.g. Longhurst et al 2008) or so-called non-representational theory, especially the more body-centric and experimental focus of Edensor (2007), McCormack (2008) or Paterson (2009), this is surprising. Tracing the path of touch within human geography in particular it is largely through the uptake of reading across disciplines, for example the influential work of ecological psychologist Gibson on the haptic system (1968), that Rodaway writes a corresponding chapter on the haptic in his *Sensuous Geographies* (1994).

Meanwhile, a niche body of related research around visual impairment, cognitive mapping and navigation aligned some geography scholars, such as Reginald Golledge and his co-workers, with wider literatures on blindness and environmental psychology (e.g. Golledge 1993). It takes another leap through the decades to see touch reappear, being loosely connected with the emergence of sensory studies within other disciplines, but also part of the ongoing operationalisation of somatocentric research.

Given both the historical origins of the geographic discipline as predominantly visual survey and cartographic display and the resurgence of interest on scholarship around the body in the 1990s it is curious that, firstly, relations between geographical scholarship, spatiality and touch remained under-explored, and secondly that rigorous attention to somatosensory experience in general was ignored for so long. While touch still remains marginal in geographical scholarship something is beginning to shift, connections between disciplinary fields interested in space and place are occurring, and particularly fertile transdisciplinary research programmes in health and well-being, in therapeutic spaces and landscapes (e.g. Bingley 2003, Butler and Parr 1999), in the new performative spaces of the body and movement through landscape, sports and tourism (e.g. Edensor 2006, Spinney 2006, Saville 2008) and elsewhere are coalescing to provide fertile ground for research in touch, haptics and the body to take root and flourish. The more recent interest within human geography with the affective aspects of everyday spaces and performance sometimes invites a specific focus on the sensual and the pre-cognitive (e.g. Thrift 2007). Much of this work moves beyond representational (visual and textual) readings of place and environmental interaction to an interpretative emphasis on emotive states and embodied practices. However, any so-called 'performative turn' in human geography has, we believe, so far underplayed the socio-cultural complexity that regulates touch in different places – the conventions of when, where and with whom one can touch. How are these conventions policed? To what degree are places of touch gendered, and how does age, culture or ability become associated with touch? To what degree do spatial contexts for activities matter (work places, retail space, domestic homes, etc.). Some of the new work presented in *Touching Space, Placing Touch* are speaking to this.

As should be evident by now, and as reflected in the chapters brought together in *Touching Space, Placing Touch*, we are moving away from the seeming immediacy of an individualised cutaneous touch, moving simultaneously 'inwards' by complicating ideas of sensations throughout the dispersed body, but also 'outwards' between bodies and subjectivities. Touch is integral to every aspect of social action and its symbols and meanings deeply infuse all cultures. It is the most intimate spatial relationship between people, and a vital and subtle communicative practice. The places where people want to touch, are allowed to, obliged to, refuse to, or are forbidden to touch form a complex and delicately-patterned socio-spatial landscape that is negotiated largely subconsciously. Children learn their place and where to touch and, importantly, not touch. Furthermore, people understand and organise the world through touch in differing ways. As Classen (2005, 1) notes '[t]ouch is not just a private act. It is a fundamental medium for the expression, experience and contestation of social values and hierarchies.'

Importantly, the role of touch is not universally positive. Our project is no simple-minded call for more, or better, touching practices. The inequalities and unevenness of tactile experience materialise and are enacted within particular places, and are accordingly processed and read within those places. We believe there is much valuable work to be done in mapping the differentiated landscapes of touch in some detail, seeking to highlight varying patterns of tactile interactions within specific places and within the conduct of particularised spatial practices. The places of touch are inevitably and sometimes powerfully experientially differentiated, and status and social role unquestionably affects how we come into contact with the spaces brought into being. A strong example of this is the difference between being a nurse or patient in a hospital, involving professional touching (cf. Andrieu et al. chapter in this volume). This offers up routes for those who want to modify space to effect progressive social change. As an illustration of this we cite the work of radical geographer William Bunge (1971) who sought to map the everyday experiential landscape of children in the Fitzgerald neighbourhood of Detroit, detailing, for example, how designated school playgrounds were haptically hostile and undesirable for children to play on. His analysis highlights the extent of jagged objects and sharp glass fragments on the playgrounds. He asks pointedly with respect to the ground-level view of a school playground displayed in Figure I.2:

> What is it that the human child in Fitzgerald actually touches? Is this a suitable surface for human contact, or is it just cheap, easy to maintain, easy to drain? Or is it deliberately inhuman so as to discourage after-school use? Would anyone want to picnic here? (Bunge 1971, 155)

Empirical Research on the Spaces of Touch

If, as already suggested, it remains impossible to presuppose any unitary conception either of 'place' or 'touch', any claims to 'know' either are indeterminate. Consequently we believe it unwise to assume any consistency in how researchers engage with such concepts empirically, or to impose any artificial classification scheme. Thinking about method has, therefore, been an illuminating part of developing *Touching Space, Placing Touch*. The social sciences as a whole have woken up to the idea that 'place matters', in particular in terms of the metaphorical and psychological dimensions and experiences of place, as opposed to rooting our understanding of place in any concrete spatial framework. A diversity of methodological approaches has arisen in human geography as a result, with a growing appreciation for, and playfulness with, the situated research encounter (cf. Hawkins 2011, MacKian 2010). In particular researchers have used storytelling, performance and visual ways of representing their empirical encounters (see, for example, Latham 2003, Laurier 1998, Pearce 2008).

Figure I.2   Tactile hostility of school playground exhibited in the critical analysis of radical geographer William Bunge

Source: Bunge 1971, 155

The nuanced socio-spatiality of tactile engagements, however, has remained less empirically interrogated. While Wylie (2002 and 2006) for example has conducted fieldwork exploring the performativity of landscapes in terms of embodied sensory experience, and explicitly acknowledges the role of the haptic in this, he is more interested in the kinetics of narrative than the place of touch. Yet, as Anne Volvey in this volume argues, touch has always been a part of what geographers do during fieldwork, but only recently has it been considered a valuable and valid source of real 'data', thereby opening up numerous questions about the place of touch at the heart of scholarly practice. Given the range of contributions in *Touching Space, Placing Touch* we might ask how the varied methodological approaches of social scientists, and human geographers in particular, could be reconfigured, adapted and extended to do greater justice to the intricacies, delicacies and contradictions of touch. Does the researcher try to measure touch itself in some way, or satisfy themselves with (often imperfect) proxies for it in the form of words and categories? Is it ethical and appropriate for academics to attend to their own haptic experiences as an undeniable part of the empirical process? As a number of the chapters in this volume begin to demonstrate, the answers to such questions are plural, but all highlight to varying degrees how attention to touch grants scholars some form of access to a subjectively-constituted interior experience and understanding not always discernible through behavioral observations. For touch lies at the interface between the perceived interiority of an embodied subject and the exteriority of the world they bring into existence through actions and relations. Several of the authors in *Touching Space, Placing Touch* engage with the modalities of touch through discursive analysis without themselves touching the places they delineate, others openly embrace the inevitability of immediate and intimate tactile encounters during their empirical investigations, reminding us that touch exists in a relational space between those touching and those being touched. If some use well established interview approaches to obtain rich empirical material, as demonstrated in Jennifer Lea's interviews with massage practitioners, others, such as Sarah Cant, Pau Obrador and Elizabeth Straughan, opt for more participatory research methods in an effort to get closer to the place of touch in their particular research contexts. This, as Hannah Macpherson says, requires a bodily and sensory immersion on the part of the researcher which, taken against the more positivist demands within social science for investigative rigour in large samples and generalisable results, sits at odds with the way academic geography has habitually been conceived. An attention to one's own embodied touch therefore – either as a substantive topic or as an element of fieldwork experience – demands that scholars take a more involved role as researchers, and reminds academics of the need to consider their auto-ethnographic role in researching, experiencing and representing tactile senses within fieldwork and research dissemination.

Touch in any context can become markedly personal and private, and as a necessary corollary to this researchers can never be certain they truly grasp the meanings and sensations of those they research. Whilst interviews can be used to probe verbalised representations of what such encounters may embody, there will always be the feeling that words alone fail to grasp the non-textual kernel of tactile experience, that that which struggles with representation strains to be articulated through language. Alternatively, by immersing themselves fully in the field of touch as participants, researchers may feel closer to what they are teasing apart, but this may be re-presented only from their own embodied position and perceived social situation. In writing about touch research, then, the danger remains that the form of embodiment assumed, the imputed body of the researcher that attempts to articulate what touch *feels* like, is a solipsistic body, an artificial horizon, an introspective abyss. The extent of rigorous research about subjective and inter-subjective touching might thereby be constrained to rich descriptive pieces of self-reflection, always mired in the local, the idiosyncratic, and unable to say anything of wider significance beyond that very personal account.

The degree of consideration given to personal reflection by several of the chapters here would suggest attention to touch throws the spotlight on the researcher as the medium through which the reader understands. Since each personal interpretation of the meaning of touch and its spatial contexts is unique, this potentially leaves researchers grappling once more with questions of representation and authenticity. In qualitative research methods more generally there always remains the possibility of falling back upon the particular words that people choose to describe or explain experiences, with the insertion of lengthy and carefully transcribed quotes in academic analysis. However, when investigating the place of touch experience, the inability of verbalised text or visual observations to truly convey embodied meaning and experience is one of the main concerns underlying the development of 'sensory studies', and necessitates an inquiry into supplementary means of, if not 'knowing' exactly, then 'reporting' that world. If what we claim is self-evident – namely that touch is about something more than the language available to describe it or the representations of it through words and images – where does that leave scholars in terms of understanding genuine and authentic experiences of touch for others? Sarah Cant's chapter in this volume left the question of gender implicit, for example, and the reader cannot ascertain whether she danced with a man or a woman. However, the reader might assume that the experience would be very different for both partners depending on the gender of each for all sorts of personal, cultural and physical reasons. Whilst this does not detract from the seductive storytelling of the embrace and the haptics of the dance, it suggests there are other routes into touch which might be attempted from researchers' own embodied positionalities.

Echoing our earlier observation that touch is not universally positive, several chapters in this volume highlight the therapeutic value of touch. It is perhaps not by chance that the subjects chosen for fieldwork involve largely non-threatening haptic experiences, often those the researcher themselves were already personally immersed within, such as recreational dancing (Sarah Cant), receiving beauty treatments (Elizabeth Straughan), or participating in countryside walking (Hannah Macpherson). Since not all touch is therapeutic, nor altogether positive, we should acknowledge that *Touching Space, Placing Touch* as a whole fails to engage with some of the more problematic aspects of touch and its associated intimacies. Whilst Jennifer Lea's massage practitioners may suggest touch can help a client 'return' to their bodies, instances of touch such as an unwanted grope will instead become the catalyst for a disruptive experience, abruptly ripping the subject from their usual habitus. Completely new and potentially uncomfortable issues then irrupt for social science researchers, both physically and ethically, and few researchers would wish to open themselves to unwelcome or potentially threatening tactile encounters. Nonetheless, such considerations suggest there is another significant terrain of touching place and placing touch unintentionally omitted from this particular collection. While no contributor volunteered this topic, it would inevitably have offered significant empirical difficulties and potentially stark provocations for considering touch otherwise, outside of the predominantly positive associations within this collection.

Despite this limitation, and given the centrality of touch to our basic human flourishing (as shown in this collection by Bernard Andrieu et al.), sense of self and identity (Amanda Bingley), and ability to relate to others (Hannah Macpherson), any methodological advances in understanding, engaging with, and explaining touch can only be welcomed. Haptic experiences of and in fieldwork are a core part of what many academics do as researchers, and how one produces knowledge, regardless of the substantive topic under consideration. In acknowledging the importance of touch we must also accept that, as Paterson (2006, 2007), Pau Obrador (this volume) and numerous others recognise, there is a close relationship between touching and feeling. To attend more fully to touch in academic research, therefore, invites scholars to attend to their feelings, potentially an uncomfortable and unfamiliar demand in many professional settings. Touching methodologies therefore prompt us as researchers to explore how *we* feel and how we feel *about* our subject matter. Viewed in this way, as Anne Volvey succinctly puts it in this collection, we work 'with' rather than 'in' the field, and this raises issues of responsibility which are rarely considered in the routinised process of 'ethics committee' clearances.

Above all, collectively the chapters in *Touching Space, Placing Touch* remind us that touch is relational, is co-produced, is co-constituted in a series of configurations between human and (non)human, and people and spaces alike. If in late capitalism the prevailing cultural and corporate tendency is, contra Hobbes' collective 'commonwealth', towards individual insularity and atomisation, we should welcome an empirical stance which approaches such intertwined and intersubjective realities. As we continue to seek opportunities for deep connection in such a world, placing empirical touch centre-stage represents a collective phenomenological 'feeling our way', or perhaps a tentative 'groping', within this emerging, exciting, haptic territory.

The Shape of Touching Space, Placing Touch

The twelve new chapters brought together in *Touching Space, Placing Touch* reflect an openness to various approaches to tactility and spatiality. The diversity of material, is a measure of the liveliness of the research going on right now in terms of both theoretical positions and methodological approaches. Indeed, within human geography this research interest continues to expand and develop, indicated for example by subsequent conference sessions such as *Touched by Geography* organised by Deborah Dixon and Elizabeth Straughan at the Association of American geographers (AAG) conference in 2009. Since 2007, as we note elsewhere in more detail, work on moving bodies, on kinaesthesia and sporting bodies has grown in interest, alongside work more recently on visceral geographies (e.g., Hayes-Conroy and Hayes-Conroy 2010) which clearly complements this area.

The chapters here are not consciously arrayed in thematic order, nor grouped into artificially imposed categories invented by the editors. While acknowledging that the contributors give a diverse set of snapshot viewpoints onto tactility and spatiality, we do see some significant commonality in terms of the thrust of their theoretical arguments. We can identify six points of intellectual intersection shared across the chapters to varying degrees. Before summarising the chapters and identifying ways in which they respectively speak to these themes, we outline the themes in general terms. Firstly, many contributors make the case that research on the geography of touch has been hidebound by the dominance of the visual register as a way of knowing the world, and awareness of the constraints of textual inscription for representation. The second point advanced effectively by several of the contributions is the use of their own bodies, and its haptic experiences, as the central 'investigative tool' for generating valid empirical observation for scholarly interpretation. Thirdly, nearly all the chapters speak to the point, albeit in heterogeneous ways, that touch necessitates a relational approach and not simply the solipsistic subjectivity assumed in thinking about touch on the skin, or individualistically-determined isolated sets of sensations. Accordingly, for some contributors, the most significant observations within their empirical analysis is the attempt to 'map' the relational spaces in-between bodies that acts of touch bring into being, in some senses, and thereby to really begin to understand what effects touch has on people's sense of the world and their place within it.

The fourth theme develops from that of the third. Rooted in the haptic, and emerging from sensual relations with others, if the sense of 'placing' our bodies through touch fosters the acknowledgement that 'touch' and 'touching' is irreducible to superficial 'surface' or cutaneous (skin) sensations, then there is a concomitant need to investigate more thoroughly how the haptic realm operates not as a single modality, as a consistent but vague visceral 'sense' in itself, but instead is neurologically constituted and consciously felt in multiple ways, often at different 'depths', or is felt as several steps removed from the bodily site or situation of actual or assumed physical contact. The focus on the multiplicity of touch therefore foregrounds a relation between an assumed interiority of the body, something reiterated throughout folk psychology, and the consciousness of touching and being touched by other bodies. The reciprocity of physical touching, the observation repeated from Husserl, Merleau-Ponty and Irigaray that touching always inevitably implies being simultaneously touched, has an affective correlate. In other words, that profound yet indirect relationship between acts of physical touching and being emotionally 'touched' finds its echo in contemporary metaphors and idiomatic expressions, where 'touching' and 'feeling' persist as metaphorical expressions of a physical act, and relates to the fifth theoretical theme that several chapters consider. This theme identifies how physical tactility helps people connect emotionally, and often in novel ways. In the cases of therapeutic spaces and healing practices this connection might be to an inner sense of selfhood assumed by the research participants, but this does not foreclose alternative conceptions of relationality between bodies and energies, nor the multitude of ways that affective and tactile practices are or could be enfolded and co-constituted in various spatial contexts. Lastly, the sixth thread that draws several of the chapters together is a focus on the work of touch in the different parts of the 'body industry', and in particular on those who make their living performing 'body work' by selling haptic labour. This type of relationship between spacing touch deals most significantly with the novel ways that the body and tactility enters the sphere of leisure-oriented capitalism, whereby the body and the haptic senses enjoy a new significance. There are new modes of somatic address that proliferate from certain sectors of the service industry that complement more traditional or established therapies and care of the body. While issues of gender pervade each of the themes to varying degrees, the uses of tactility and the body within therapeutic spaces, and the asymmetrically gendered deployment of haptic labour within the body industry, entail that gendered labour and power relations are more pointed in this regard.

Jennifer Lea's work on the place of touch has focused on the spaces of health and care giving, especially in relation to the practices and meanings around therapeutic massage. Her chapter, *Negotiating therapeutic touch*, focuses on extending existing scholarship around the political economy of 'body work' (such as nursing, social care, beauty treatments) by conferring more concern for the phenomenological experience of the body actually doing the work. Her analysis draws upon the influential writing of Michel Serres (2008) which offers a complex, and therefore perhaps a more comprehensive, reading of the body in terms of corporeal sensations and most significantly the ways these emerge through social relations. Serres' ideas, according to Lea's interpretation, attend to the ways that touch can reach through the surface of the body and connect across bodies, admixing being and world in ongoing and unpredictable relations. This notion of 'mixing' effectively complicates the easy trap of inside/outside dualism around the analysis of embodied touch, where sensations are not created simply at the skin boundaries but arise from the relations. The body therefore is not a meaningfully bounded entity. The significance of sensory relations for understanding the world needs to be addressed, for this is how we become: through feeling the world.

Jennifer Lea argues that Serres' ideas of fleshy and feeling bodies, and his concepts of 'mixing' in particular, can help social scientists conceptualise the places of touch in novel ways and begin to reveal how bodies are constituted in relation to their spatial context. Considering how the body is not a bounded, unitary object, but emergent through relations including sensory ones, using in-depth interviews she analyses the working touch of therapeutic massage practitioners. Her analysis shows how language – in the form of verbal accounts of clients given when undergoing massage treatment – is insufficient to explain their bodies and expose the 'problems' residing within. The massage practitioners, in the act of 'mixing' their bodies with the clients', can in some way connect to a kind of demonstrable 'truth' through their skilled touch that is inexpressible through words. Yet, as Lea details, such truth when activated through touch comes laden with tensions. These tensions can arise from the real difficulty some clients have in dealing with the impact of a truth 'exposed' to their consciousness through the touch of massage. Furthermore, there can be issues in coping with ongoing emotional problems that emerge through mixing bodies because the place of touch is bounded by the time and space of the treatment room and constrained by the business relation that exists between client and therapist (and their profit derived by offering their haptic labour). The mixing of bodies in this way is problematic as it is part of 'body work', and cannot be freely expressed as an open-ended care-giving relationship.

Pau Obrador engages with ideas around the place of touch in the context of tourist activity, focusing on the intellectual gap in academic studies of tourism in understanding the sensuality of material practices. His chapter, *Touching the beach*, makes a case for the significance of bodies, their corporeality and the sensual nature of encounters with spaces, to extend tourism studies beyond its conventional analytical concern with the optical senses and visual culture exemplified by the centrality of the gaze metaphor to much literature on tourism. A specific focus on the haptic basis of tourist activities, Obrador argues, can help to understand the power of their sensual experiences, pleasurable feelings and playful meanings, and thereby move scholarship beyond the constraints put up by the conventional focus on the predominantly visual consumption of places.

Drawing on earlier empirical material, Obrador's spatial context is the beach, a site essential to much of the tourist industry, renowned as a space offering a distinctive ensemble of materialities – the famed three 'Ss' of sand, sea and sun. The beach is also lionised as a public space for playful encounters in contemporary Western culture. As such beaches are interesting tactile spaces to investigate and, according to Obrador, their range of sensual opportunities are surprisingly under-researched by tourist studies. Given the ways bodies can have multiple, and quite often distinctive, sensual encounters at the beach, as an essential part of its broad ludic appeal – freedom to get onto the ground and play in the sand, laying flat out soaking up the sun, splashing about in the sea – clearly the beach cannot be fully explained by documenting the visual register alone.

Obrador looks to understand the beach by an engagement with the tactile appeal of sand, sea and sun, and deploys three distinct conceptual tools to achieve this. Firstly, he shows how a focus on the complex modalities of touch can overcome the isolated viewpoint, which are inherent in ocularcentric approaches. His concern for looking beyond vision to use the tactile senses to unlock experiences of place speaks directly with Anne Volvey's call in chapter 5 for attention to the more-than visual fieldwork practices in geographical scholarship. Obrador's readiness to engage with the embodied reality of sensual experiences of play and pleasure also correlates with other recreational tactilities such as those documented in Jamie Lorimer's chapter on tourist encounters with wild animals and Hannah Macpherson's discussion of practices of guided walking holidays. Secondly, Obrador conceptualises an expanded realm for the haptic, one that extends touch well beyond the bio-psychological feelings received at the fingertips. As he shows the sensual encounters with the plasticity of sand, the heating rays of the sun and the enveloping motion of the sea can only be meaningfully read with a 'thick' haptic perspective that registers the many ways touch occurs upon, within and across our bodies. The notion that touch is more than skin deep is a thematic current in several other chapters, including Jennifer Lea's discussion of the depth of feelings released through therapeutic massage. Thirdly, he conceptualises the value of touch for its capacity to bring forth the textures of places and highlight the feeling of authenticity that comes through the haptic sensorium.

Being able to describe more fully the textures that make places unique is valuable for materialistic approaches in social sciences. The focus on materialities is shared with Martin Dodge and Rob Kitchin examining the changing tactilities of public toilets arising from new technologies. Furthermore, the analysis Obrador presents in *Touching the beach* draws upon first-hand insights from participant observations on beaches and interviews with beachgoers, along with some insightful and introspective use of his own bodily experiences. In this regard Obrador's empirical approach has clear methodological parallels with work presented in this volume by Sarah Cant on bodily experiences of tango dancing and Elizabeth Straughan on receiving beauty treatments.

The significance of touch in art therapy is explored in Amanda Bingley's chapter where she draws on influential ideas from psychoanalysis such as Winnicott (1971) to explore the notion that the tactile senses are a pragmatic alternative for articulating more than can be verbalised within a therapeutic context. Not only non-verbal communication, but more-than-verbal expression, Bingley believes, is the essence of why art therapy can be so effective in helping people recover from trauma or cope with serious and terminal illnesses. As she shows in *Touching space in hurt and healing*, haptic mechanics in the body can make certain kinds of 'deep' connections to the brain to unlock emotions and memories in ways that other senses cannot.

Bingley's empirical context is creative play and art making which is now widely deployed in the healing practices across many different institutional settings. Art therapy 'works' because touch is the deepest sense, the least deceptive, the most difficult to fake or be fooled by. This is because the haptic realm forms the 'ground' upon which everything else is built, it comes first to us as we form in utero and it matters the most throughout life. Tactile art activities in therapy sessions can simultaneously bring somatic responses and emotional feelings to the surface, to re-present within consciousness, helping to reconnect them with their body, which is often the 'problem' that a person must confront in their healing. Here Bingley's analysis articulates similar ideas to those advanced by Anne Volvey on the elemental nature of touch and how it works to unlock new knowledge about the self. The purposeful exploration of inner selfhood through the medium of touch is also a central notion in the discussion by Sara MacKian on the significance of spiritual encounters, Jennifer Lea's contribution on therapeutic massage and Jamie Lorimer's discussion of people 'finding themselves' through tactile encounters with elephants on volunteer holidays.

Bingley develops Winnicott's notion that touch works within a domain, the 'potential space', that lies at the interface between perceived Self and observed Other, inner being and exterior world. But for Bingley *what* one is touching matters, as some objects simply work better as a tactile medium. They are not just bare materialities but active agents within this 'potential space', enrolling bodies into the world. She looks in particular at the materialities of clay and sand and their tactile effectiveness in art therapy practices, detailing ways these materials provide naturalistic, workable forms of physical expression of people's inner, imaginary, world, through the intensely sensitive feelings of hand and fingers. Such materials can be made to 'talk' without conscious effort, the inner voice can 'speak' to them through the subtle, elusive power of touch. Yet Bingley argues from her wide reading of the applied literature on art therapy that the fundamental significance of the tactile is all too often eluded. She delineates how the act of touching, what it can do for individuals seeking healing, and how it does it, are rarely acknowledged. Possible reasons for this include the primacy of the visual in recorded research, the unspoken hierarchy of senses that tends to diminish the significance of the haptic realm, and the fact that touch forms a fundamental 'ground' for everyday perception means that often it is dismissed as mere background, and as such is hard to articulate through text. This argument has much in common other chapters on the need to look beyond the visual register including Pau Obrador's analysis of the beach, Jamie Lorimer's examination of touching environmentalism, and more generally Anne Volvey's call for new forms of fieldwork attendant to the tactile.

Elizabeth Straughan's chapter is concerned with the 'body industry' and the everyday place of therapeutic touch as conducted in high street beauty salons. She is particularly concerned with understanding the nature of facial treatments in terms of the emotional exchanges and tactile relations between the touch giver, the beautician, and the touch receiver, the paying client. This 'body work' involves a significant amount of haptic labour with a range of tactile practices, coming from the key site of the hands of the beautician and directed to one of the most sensitive parts of the body, the face. Delivering a distinct form of therapeutic touch which will improve the (perceived) appearance of the client's skin, the emotional labour of the beautician's job should not be underestimated. Success in this regard, as Straughan shows in *Facing touch in the beauty salon*, is as much concerned with psychological connections as it is with the physiological condition of the skin.

Empirically Straughan's work is focused on the feminised nature of the salon space and draws upon her own experiences of receiving facial treatments, recounting the routines and the staged ambiance of the place, the verbal discussions with the beauticians and the variety of feelings engendered within her, both at a somatic and an emotional level. To provide an explanatory interpretation of her own embodied experiences of receiving therapeutic touch in beauty salons, Straughan seeks to delineate the contradictory process of generating a sense of relaxation under soothing hands whilst also unleashing anxieties about this very process, since skin 'problems' are necessarily being exposed to scrutiny in order to be 'treated'. She brings to the fore Katherine Hayles' (1997) theoretical concept of 'corporeal anxiety' to account for the dialectical nature of beautification: that it depends on the decaying nature of our bodies and simultaneously the desire to counter this decay. The result is a compulsion to seek external 'solutions' that promise the restoration of something 'lost', now centred in a commercial salon context that proffers scientifically labelled products with their restorative and regenerative properties. Straughan's analysis of therapeutic touch and the anxieties provoked in terms of corporeality and the effectiveness of commercial performances around 'body work' and emotional labour, resonates strongly with Jennifer Lea's work on massage practitioners and Andrieu et al.'s chapter on research regarding the tactile work of doctors and nursing staff in medical settings.

Straughan's analysis of the nature of bodies suffering corporeal anxiety draws upon two elemental ideas from Luce Irigaray's (1993) work. Firstly, she asks us to think about the nature of bodies in the salon using the notion of '*morphé*', which acknowledges the temporality of the body, continually coming into being and breaking apart, requiring ongoing care. Secondly, Straughan uses Irigaray's notion of 'porosity' to account for the body's vulnerability to fragmentation and yet its fundamental openness to treatment. Bodies are not closed entities, and the porosity of the beautician's hands and the face of the client is a factor in facilitating 'good' or effective therapeutic touch. Along with Rachel Colls' contribution in this volume looking at the tactile fatness of women's bodies, and Sarah Cant's discussion of embracing tango dancers performing fluid movement, Straughan's theoretical engagement develops Irigaray's work on bodily difference and subjectivity for placing and conceptualising touch.

Anne Volvey's chapter charts the shifting 'knowledge regimes' that she argues underpin empirical, fieldwork-based, geographical scholarship. Such regimes have an often unacknowledged *episteme* that is critical in guiding how geographers come to know what they know about the world. As others have done, she contends that much traditional geographical fieldwork has been based upon the primacy of the visual survey as a way of knowing the world and has been criticised for its inherent masculine biases. Addressing this, Volvey delineates the emergence of an alternative knowledge regime for fieldwork, again spurred on in part by feminist geographers, one that is qualitative and based on discursive ways of knowing. While Volvey acknowledges the positive potential of this 'qualitative turn' in fieldwork she is simultaneously critical of this knowledge regime, centred as it is around 'talking' to subjects, and its tendency to overlook the other embodied senses, particularly the innate haptic registers of the researchers themselves. Consequently, in her chapter *Fieldwork: how to get in(to) touch*, Volvey calls for an 'expanded' knowledge regime for fieldwork that encompasses not just what is seen or said, but also how scholars *feel* about the world. This entails an ambition to accept the tactile experience of researchers as a valid form of data that should be treated as seriously as any other source in the production of scientific information.

For a truly haptic 'knowledge regime' Volvey deploys ideas from psychotherapy around 'transitionality', focussing our attention on the significance of the 'between-ness' of bodies and the world, the interface of 'me' and 'not-me', as a central part of our being. Using the ideas of 'transitionality', derived partly from Winnicottian psychotherapy, as a way to think about the significance of the haptic realm to self-being, clearly shares a sympathetic conceptual background with Amanda Bingley's chapter on touch in art therapy. Volvey argues fieldwork should be seen as a transitional practice, one that brings to the fore the unconscious incorporation of all manner of 'data' from feeling (as opposed, presumably, to the more usual 'surveying') the field. Thus, fieldwork should not be seen in terms of an external agent collecting material 'in' the field, but more an experience performed 'with' the field. It enhances fieldwork, giving substance to non-visual experience and also more-than verbal explanations, enhancing the repertoire of social science scholarship. To some extent this effectively demonstrates comparable approaches in this volume (especially Hannah Macpherson and Pau Obrador), where the research builds upon empirics felt by the researchers' own bodies. Volvey argues this agenda extends feminist approaches to fieldwork by moving towards an *episteme* purposefully centred on more embodied, corporeal feelings. This kind of haptic knowledge regime, coming out of fieldwork-as-withness, potentially opens up notions of the researchers' own 'sense of self', allowing them to explore their existential motives as well as their basic socio-political positionality. This could move geographical scholarship forward in some surprising directions.

Hannah Macpherson's chapter describes the activity of guiding blind and visually impaired people in the context of recreational countryside walking. This leisure context provides a revealing and singular context to consider relations between embodiment and touch. Her work seeks to understand what it means for two people to 'move as one' through the landscape. Drawing on qualitative fieldwork in which she was a volunteer guide on holidays with blind and visually impaired people, Macpherson's analysis demonstrates the value of 'sensuous ethnographic observations' in combination with interviews and photography. In her chapter *Guiding visually impaired walking groups*, Macpherson deploys her own guiding experiences to delineate the 'in-betweenness' of the touching bodies, of guide and follower, and highlights how subtle yet significant these inter-corporeal spaces are. Such inter-corporeal spaces are all too easily overlooked in geographical scholarship, but especially in much conventional individual-centred 'wayfinding' research with visually impaired people. Thus her research has wider value in highlighting the nature of the spaces and practices where bodies come together experientially, beckoning the intercorporeal world into some kind of symmetry, if only in fleeting and partial ways. Such temporary inter-corporeality through touch is shared to some extent with Sarah Cant's analysis of the social embrace and paired movements of tango dancers, and also in Martin Dodge and Rob Kitchin's consideration of use of public toilets that are shared with strangers.

Macpherson develops this notion of 'inter-corporeality', building upon Merleau-Ponty's (1962) well known idea of 'body schema' and how they are extensible beyond the strict physical bounds of skin and bone. Her analysis of guided walking in the countryside illustrates how body schemas can become coupled, with two bodies exhibiting kinetic synchrony, like a dance along the rural footpaths and mountain trails rather than a mechanical follow-my-leader march. In trying to understand how coupled body schemas work Macpherson deploys an ethical perspective, thinking of touch as a gift that is given and received. In this inter-corporeality, importantly, the visually impaired person is not a passive vessel or pitiable recipient of charity, but is instead empowered by surrendering their independence and by choosing to gift their trust to the guide. The guide's body does not give help as such, but receives this gift of trust and has responsibility for its care. The gift of trust is made real in the incorporeal space of touch. We can see parallels in this exchange of trust within an inter-corporeal space of touch with Jennifer Lea's chapter on 'mixing bodies' in therapeutic massage and Elizabeth Straughan's chapter on the touch between client and beautician in the gift of facial treatment.

The chapter by Bernard Andrieu, Anne-Flore Laloë and Alexandre Klein examine touch in the context of medical models of the body and health spaces, deploying a range of conceptual cases studies to shed light on what they see as new kinds of 'biosubjective care'. It shares the focus on touch in therapeutic spaces and practices by Jennifer Lea and Amanda Bingley in this volume, but concentrates on the relation between medical models of the body and strategies of healthcare. They argue that the caring potential of empathetic touch for therapeutic purposes has been consciously disregarded in mainstream, science-centric, masculinist, medicine, and has also been elided in conventional modes of nursing that are focused on material hygiene and the orderly management of ill bodies. Healing touch has been relegated to the margins by trained health professionals, and stereotyped (even stigmatised) as 'alternative' medicine. In *Touch, skin cultures and the space of medicine,* Andrieu, Laloë and Klein seek to explain this through the dispersed ontological status of the body in scholarly knowledge. They discuss the tendency to divide ways of understanding, with the 'lived body' on one side with its concern for surgical dissection of tissues and organs, differentiated from the 'living of the body' and its critical questioning of representations and discourses. Andrieu et al. contend that a focus on touch in terms of biosubjective care could be a useful epistemological tool for investigating the body, fusing together bio-physical and psycho-social bodily models.

Furthermore, Andrieu et al. aim to counter the particular kind of professionalisation of touch within medical practice and thereby to rethink medicine as a 'tale of the skin', one that acknowledges and integrates diverse tactile therapies and their expanded and enhanced capacity to heal patients. In this sense their chapter has resonances with the holistic arguments around touch and the body advanced by Amanda Bingley's chapter in her consideration of tactile art for healing, the commercially determined 'body work' discussed by Elizabeth Straughan, and the professionalisation of kinds of healing touching within the beauty industry by Jennifer Lea.

Jamie Lorimer's chapter takes us in quite a different direction and into a distinctive empirical context, concerned as he is with the touching that marks interspecies relations, in particular the complex patterns of embodied and non-verbal interactions that occur between humans and companion animals within recreational spaces. Drawing upon his fieldwork, Lorimer speaks of a 'touching environmentalism' which exploits the capacity to be in touch with, and to have actual tactile encounters, especially with charismatic animals such as koala bears or dolphins, and has become central to the wildlife conservation strategies of NGOs and ecotourism companies. As he discusses, there is an inherent 'captivity paradox' here, whereby wild animals are kept under close control so that paying visitors can get close to them, yet also have supposedly 'authentic' encounters with them.

In *Touching environmentalism,* Lorimer investigates the triangular relationship between an aging elephant, a Western woman on a volunteer holiday, and the local *mahout* who cares for the animal. Using empirical material gathered at a Sri Lankan elephant sanctuary through participant observations, filming and interviews he considers the different kinds of feelings generated, and possible meanings in, episodes of touching in staged encounters. In this context, Lorimer also exposes conflicts around how interspecies encounters should proceed and the different models of welfare, care and cruelty are at play in the sanctuary situation. How should human-elephant contact be considered? The elephant is ambiguously positioned: neither a wild beast nor a subservient pet. Instead, the elephant exists within a web of relations and knowledges as what Donna Haraway (2008) terms 'companion species', with long histories of relationships with humans.

By making reference to influential ideas from Haraway's work, Lorimer proceeds to think about the deeper histories and troubling relations that mark many kinds of human exploitation of animals, and how contemporary responsibilities might be renegotiated. For this he delineates various modalities of touch and identifies how these vary across different actors (involving asymmetrical power relations and often incompatible embodied knowledges), along with attention to their discrete historical antecedences. In particular he explains how pre-existing colonialist notions, centred around particular kinds of appropriate and inappropriate human-animal relations, is still significant to contemporary touching environmentalism. He charts how the elephant in particular has been perceived as an exotic icon because of its size and strangely compelling physiology, sought after by hunters and subject to the deadly touch of their rifle bullets, and also subjected to the scopic curiosity of nascent natural historians. This attention has more recently morphed as feminised conservation agendas have shifted the *episteme* from masculinist knowing (derived from elevated visual observation) to a closer, more embodied and literal way of getting in touch with animals.

Lorimer shows how such connections can be made to heighten interest in the therapeutic potential (for the human) of the tactile sensations within human-animal encounters. In many of these encounters people recount being touched, in some spiritual dimension, through their communing with 'pure' nature by the supposedly unmediated act of touching sentient animals. This has become another of those packaged experiences for affluent consumers, such as swimming with dolphins, as part of the wider growth of the postmodern 'experience economy'. It is certainly the case that those paying to work as a volunteer with elephants in the Sri Lankan sanctuary sought some spiritual self-healing from their physical contact with the animals. In making the conceptual connection between the immediacy of tactile encounters and communing with something much larger, an other that awakens larger spiritual feelings, Lorimer's chapter shares some corresponding concerns with Sara MacKian's chapter by looking at the significance of tactile connections to spirit for everyday well-being.

The importance of tactile engagement within everyday environments is the subject of the chapter by Martin Dodge and Rob Kitchin. While their techno-centric empirical approach diverges from other contributions, they seek similarly to understand something of the social meaning of socio-spatial tactile interactions. They also demonstrate the potential to mediate and modulate the haptic landscape in contemporary practices of technologies such as computers, software algorithms and digital sensors. *Towards touch-free spaces* provides a preliminary analysis focused on the mundane but overlooked space of the shared public toilet. Examining the ways that sensors and software are deployed to automate bathroom fixtures and fittings, key aspects of toileting practice can proceed without the need for direct hand touch. Their contribution draws upon anthropological ideas around the emotive power of disgust and the cultural categorisation of dirt using Mary Douglas' (1966) influential notion of 'matter out of place'. Toilets are inherently dirty places, and the desire to control disgust by reducing points of potential contact and contamination is strong. In the context of shared public toilets, the problem of surfaces that have been touched and thereby contaminated by strangers' bodies becomes prominent. The analysis advanced by Dodge and Kitchin speaks to the relational significance of touch within the quotidian space of the bathroom, the apprehension and anticipation of the (now absent yet, evidenced by dirt and detritus, somewhat materially present) body of the other, and complements the discussions of the importance of inter-personal tactility in therapeutic spaces (see Jennifer Lea and Elizabeth Straughan chapters) and recreational activity including Sarah Cant's dissection of tango dancing. Much of Dodge and Kitchin's chapter audits the range of touch-free technologies that have become available from bathroom manufacturers and the kinds of promotional discourses expounded to sell these products, including the central claim around their hygienic capacity, the offer of efficiency and the promise of control. Looking at how touch-free technologies are being sold is bound into wider capitalist enterprises that exploit tactile landscapes and haptic labour for profit, something that is well explored in a number of other chapters including Bernard Andrieu et al.'s analysis of 'biosubjective care'. A key conclusion of their chapter is to highlight the current real-world failure of technologies to meaningfully produce consistently touch-free spaces, that accordingly deliver to users a way to control these contaminatory contacts and therefore their disgust. In terms of a broader consideration of wholesale attempts to automate the landscape as 'intelligent environment' and 'smart spaces' using software technologies, the central question posed by this work concerns the feasibility and, crucially, desirability of the removal of so many routine tactile encounters within material space. Questioning the marketing and managerial rhetorics that desire to engineer away touch is important, because these are misguided in their utilitarian reading of the haptic realm, and fail to understand that touch is comprised of much more than surface sensations, a point consistently raised in other chapters.

Sarah Cant's chapter considers the significant role of touch, but also 'listening', in the experience of dancers of Argentine tango, a popular social activity around the world. The inherently improvised nature of tango, with its small repertoire of moves, handholds and forms of embrace means the dancing couple must be attuned to each other's bodies, bringing shared kinesthetic experience into inter-corporeal being. In its popular image tango is bound up with strongly gendered roles of the male dancer leading a female partner, but Cant uses this form of dancing to think more subtly and with more suppleness about difference and subjectivity by using ideas from Irigaray. In drawing upon her own experiences of tango dancing, Cant seeks to challenge simplistic male-female binaries, deploying Irigaray's notion of 'listening' across the silent shared spaces of touching bodies in the dancing embrace without reducing these to determined gendered differences. Social dances, like tango, with their varying degrees of touch-in-the-moment, are therefore interesting places to explore some routes to post-patriarchical discourses that are focused on how bodies multifariously and fleetingly 'fit' together physically and symbolically. In the *milonga* context in the UK that Cant examines, people are often dancing with strangers and the tactile intimacy of embrace needed for real 'listening' to the partner is a challenge, requiring an inner confidence to open oneself to another person, to be sufficiently 'in touch' to let the dance flow. As Cant reveals from her own experience, it does not always work. The touching embrace that links two bodies may not necessarily form more emergent connections and may simply reassert traditional follower-leader roles, because one side is not 'listening' to the other. Her analysis also shows how successful social dancing relies on more-than-physical tactile sensations exchanged between paired bodies, and in tango it is the emotional connection growing out of the tactility of embrace, but not defined solely by the 'haptic', that works because it breaks down normative passive/active gendered roles. Holding in the embrace should not be a competition, a struggle for control, but 'two singularities respectful of the other's difference'. It is not about 'equalising' that difference somehow but about 'listening' to each other through touch. Dancing bodies are replete with sensations, and the magnitude of emotional connection cannot be equated to the tactile closeness of the embrace. Sometimes light touching in open embrace facilitates greater 'listening' and engenders a stronger sensual bond between dancing bodies. The physical extent of touch is less important than the degree to which it communicates trust and a willingness to anticipate and take risks together in bodily movements. In this respect there are substantive correspondences here to Hannah Macpherson's chapter, while in Cant's engagement with, and development of, Irigrary's corpus, there is an overt overlap with the contributions by Rachel Colls and Elizabeth Straughan, who provide a reading of touch relations based on Irigaray's psychoanalytically influenced feminist philosophy that aims to rework those simplistic inherited categories of bodily being based on male/female binaries, and instead articulate alternatives expressed through linguistic constructions and metaphor celebrating bodily differences.

Rachel Colls' chapter uses paintings of self-proclaimed 'fat' female bodies by artist Jenny Saville as a way to think about particular modalities of touch and to challenge what she sees as the underlying masculinist readings of sensations emanating from Merleau-Ponty's influential writing in the mid twentieth century. Like Sarah Cant and Elizabeth Straughan, Colls' analysis in *Intra-body touching and the over life-sized paintings of Jenny Saville* draws upon the critical feminist theories of Irigaray, as well as recent 'visceral' scholarship attending to the very fleshy nature of bodily sensations and their social relations. Colls' intellectual agenda centres upon an Irigarayan understanding of touch as a relation roughly between interiority and exteriority, the materiality of fleshly bodies and world, extending this by arguing for the need to consider the particularities of size and sex of the bodies involved. Most significantly she wants to highlight how bodies touch themselves (what she terms 'intra-body touching') and the fact that size *does* matter – hence her direct political call to look at fat bodies in new a light. Again employing and developing terms from Irigaray, in this case Colls focuses upon the notion of 'morpho-logic' for female bodies centred on Irigaray's metaphor of the fluidity of mucous, as opposed to the solidity of the phallus. This highlights the distinctive nature of female bodies and how they touch – something overlooked in Merleau-Ponty's work – and reminds us that not all touching is visible and therefore available for objectification by an inquiring and acquiring gaze. Colls' chapter provides an insightful visual analysis through a reading of two politically-loaded paintings of fleshy bodies. She proceeds in this fashion firstly for pragmatic reasons, arguing that Saville's painting envisions a reality of how bodies touch themselves, but secondly she asserts that the analysis of such artworks is valuable in making sense of 'how bodies, things and matter relate "with" and "to" each other', granting evidence of touching relations residing in mucous which are not usually visible to external scrutiny, and also confronting historicised masculinist conventions in the visualisation of the female nude. This opens up an approach that challenges the normative view of fat bodies as estranged from the self and stereotypically read as socially disgusting. Rendering visceral bodies as visible in this way is potent, blurring simple boundaries between self and other, as folds of fat press against part of that same body – another quite literal way of bodies simultaneously touching and being touched.

Sara MacKian's chapter completes *Touching Space, Placing Touch* and provides an innovative dual reading of the tactile as both physical contact and psychical connection. She points to the mass acceptance of spiritual beliefs in the global North that constitute significant aspects of everyday life for many, and yet the majority may not participate in organised religious practices or churchgoing. This is most evident in self-adapted spiritualities, such as crystal healing and tarot readings, that operate beyond the institutional governmentalities of church or temple. Given this massive undercurrent of everyday alternative spiritualities her analysis highlights the therapeutic nature of spiritual touch, which need not be physically haptic to have real healing effects. In this regard there is commonality between MacKian's focus on spirituality and other kinds of 'alternative' self-help through tactile engagement discussed in other chapters in this volume, such as Amanda Bingley's consideration of art therapy and Jamie Lorimer's look at the healing that can come from getting 'in touch with' those larger natural (and supernatural) forces of nature, where swimming with dolphins is widely described as spiritually uplifting.

If the presence of spirit is unacknowledged and overlooked in much academic analysis, MacKian argues there needs to be consideration of how the more-than-bodily sensations that many people seem to genuinely experience influence their behaviours, and thereby contribute to the larger social structures and the material forms of contemporary culture. In *Touched by spirit*, MacKian looks beyond the physical structures, material practices and visual iconicity of conventional religions, to engage with the pervasive influence of everyday spiritualities. MacKian's chapter discusses how the 'lens of touch' affords a potentially valuable way to gain more nuanced sociological understandings of 'how' and 'where' spirit makes a difference within the unfolding practices of modern living. She examines in detail a range of 'points of contact' with spirit by speaking to spiritual practitioners, recounting the sensations of tactile encounters, and also considering the deeper meanings people derive from being metaphorically 'in touch with' otherworldliness in the everyday. Often extended (and non-physical) senses of touch are realised (and literally made real) through the use of specific material objects in rituals, like the white feather mentioned by several spiritual practitioners. Here bodily contact with such special objects works as a bridging device, opening connections or channels in a tangible and immediately comprehensible way between material world and immaterial spirit. This radically relational nature of touch, and the ways tactile experiences seem to bridge across and between bodies in physical and spiritual realms, speaks to the concerns of inter-corporeality and being in touch with far larger forces echoed elsewhere in this volume. MacKian's rich descriptions of the multiple ways that people are 'touched' by spirit can be taken as a progressive call for more inclusive social science scholarship. Such work not only acknowledges the importance of touch in the place of spirit, but solidifies the observation that touch, in its multifarious forms, impacts upon us in everyday life, yet is so rarely seriously considered within the social sciences.

Placing Touch

In *Touching Space, Placing Touch* each author with their adopted methodological framework contributes valuable points of reference for our growing mapping of the topographies of touch. The future of empirical research in the cross-disciplinary field of 'sensory studies' will have to feel its way, sometimes gropingly and imprecisely, in some cases using the traditional tools at its disposal, and revisiting the implications of some well-worn debates around issues like power, gender and positionality, before scholars can feel more fully at ease with writing about knowing the places of touch.

There are numerous ways in which we might locate, dissect and understand touch. The chapters collected together here range from those with a specific focus on methods of touch, to those exploring methodologies for understanding touch, and represent just the beginning. We are in the early stages of encountering, mapping and negotiating this particular territory within the academy. The project of *Touching Space, Placing Touch* therefore entreats scholars to consider, in *all* their empirical investigations – regardless of whether there is a substantive focus on touch – how their methods touch, and thereby alter, the worlds they are investigating. Do they want their touches to move those worlds, and can they avoid such touches even if they do not? The contributors in this volume go some way to exploring these questions about the methods of touch and the touch of method, thereby stimulating readers to consider and reflect upon the touching, feeling or haptic influences of their own research encounters.

References

Augustine of Hippo. 1950. *The Greatness of the Soul and The Teacher*, trans. J. Colleran. Westminster, MD: Newman Press.
Bingley, A. 2003. In here and out there: Sensations between Self and landscape. *Social & Cultural Geography*, 4(3), 329–45.

Bunge, W. 1971. *Fitzgerald: Geography of a Revolution*. Cambridge, MA: Schenkman Publishing Company.

Butler, R. and Parr, H. 1999. *Mind and Body Spaces: Geographies of Illness, Impairment and Disability*. London: Psychology Press.

Classen, C. 2005. *The Book of Touch*. Oxford: Berg.

Crang, M. 2003. Qualitative methods: touchy, feely, look-see? *Progress in Human Geography*, 27(4), 494–504.

Crary, J. 1990. *Techniques of The Observer: On Vision & Modernity in the 19th Century*. London: MIT Press.

Crary, J. 1999. *Suspensions of Perception: Attention, Spectacle and Modern Culture*. London: MIT Press.

Danius, S. 2002. *The Senses of Modernism: Technology, Perception and Aesthetics*. London: Cornell University Press.

Dixon, D.P. and Straughan, E.R. 2010. Geographies of touch/touched by geography. *Geography Compass*, 4(5), 449–59.

Douglas, M. 1966. *Purity and Danger: An Analysis of the Concepts of Pollution and Taboo*. London: Routledge.

Edensor, T. 2006. Sensing tourism, in *Travels in Paradox: Remapping Tourism*, edited by C. Minca and T. Oakes. Boulder, CO: Rowman and Littlefield.

Edensor, T. 2007. Sensing the ruin. *Senses & Society*, 2(2), 217–32.

Geurts, K.L. 2002. *Culture and the Senses: Embodiment, Identity, and Well-being in an African Community*. Berkeley, CA: University of California Press.

Gibson, J.J. 1968. *The Senses Considered as Perceptual Systems*. London: George Allan & Unwin.

Golledge, R.D. 1993. Geography and the disabled: A survey with special reference to vision impaired and blind populations. *Transactions of the Institute of British Geographers* 18(1), 63–85.

Haraway, D.J. 2008. *When Species Meet*. Minneapolis, MN: University of Minnesota Press.

Harvey, E.D. 2003. *Sensible Flesh: On Touch in Early Modern Culture*. Philadelphia, PA: University of Pennsylvania Press.

Harvey, E.D. 2011. The portal of touch. *American Historical Review*, 116(2), 385–400.

Hayles, K. 1997. Corporeal anxiety in *Dictionary of the Khazars*: What books talk about in the late age of print when they talk about losing their bodies. *Modern Fiction Studies,* 43(3), 800–20.

Hayes-Conroy, J. and Hayes-Conroy, A. 2010. Visceral geographies: Mattering, relating, and defying. *Geography Compass*, 4(9), 1273–83.

Hawkins, H. 2011. Dialogues and doings: Sketching the relationships between geography and art. *Geography Compass*, 5(7), 464–78.

Hobbes, T. 1962 [1651]. *Leviathan*, trans. W. G. Pogson-Smith. Oxford: Clarendon Press.

Hockey, J. 2006. Sensing the run: the senses and distance running. *Senses and Society*, 1(2), 183–202.

Howes, D. 2003. *Sensual Relations: Engaging the Senses in Culture and Social Theory*. Ann Arbor, MI: University of Michigan Press.

Irigaray, L. 1993. *An Ethics of Sexual Difference*. Translated by C. Burke and G.C. Gill. Ithaca, NY: Cornell University Press.

Jay, M. 1994. *Downcast Eyes: The Denigration of Vision in Twentieth Century French Thought*. Berkeley, CA: University of California Press.

Jones, C.A. 2007. *Sensorium: Embodied Experience, Technology, and Contemporary Art*. London: MIT Press.

Jütte, R. 2005. *A History of the Senses: From Antiquity to Cyberspace*. Cambridge: Polity Press.

Latham, A. 2003. Research, performance, and doing human geography: some reflections on the diary-photograph, diary-interview method. *Environment and Planning A*, 35(11), 1993–2017.

Laurier, E. 1998. Geographies of talk: "Max left a message for you". *Area*, 31(1), 36–45.

Law, L. 2001. Home cooking: Filipino women and geographies of the senses in Hong Kong. *Cultural Geographies,* 83, 264–83.

Locke, J. 1975 [1690]. *An Essay Concerning Human Understanding*, edited by P.H. Nidditch. Oxford: Clarendon Press.

Longhurst, R., Ho, E. and Johnston, L. 2008. 'Using 'the body' as an 'instrument of research': kimch'i and pavlova. *Area,* 40(2), 208–17.

Longhurst, R., Johnston, L. and Ho, E. 2009. A visceral approach: cooking 'at home' with migrant women in Hamilton, New Zealand. *Transactions of the Institute of British Geographers,* 34, 333–45.

Lowenstein, O. 1966. *The Senses*. Harmondsworth, England: Penguin Books.

MacKian, S. 2010. The art of geographic interpretation in *The Handbook of Qualitative Geography*, edited by D. Delyser, S. Herbert, S. Aitken, M. Crang and L. McDowell. London: Sage.

Marks, L.U. 2002. *Touch: Sensuous Theory and Multisensory Media*. Minneapolis, MN: University of Minnesota Press.

McCormack, D.P. 2008. Geographies for moving bodies: Thinking, dancing, spaces. *Geography Compass*, 2(6), 1822–36.

Merleau-Ponty, M. 1962. *Phenomenology of Perception*. London: Routledge.

Mitchell, W.J.T. 2002. Showing seeing. *Journal of Visual Culture*, 1(2), 165–81.

Pallasmaa, J. 2005. *The Eyes of the Skin: Architecture and the Senses*. London: Academy Editions.

Paterson, M. 2006. Affecting touch: towards a 'felt' phenomenology of therapeutic touch, in *Emotional Geographies*, edited by J. Davidson, M. Smith and L. Bondi. Aldershot, England: Ashgate.

Paterson, M. 2007. *The Senses of Touch: Haptics, Affects and Technologies*. Oxford: Berg.

Paterson, M. 2008. Review Essay: Charting the return to the senses. *Environment and Planning D*, 26: 563–569.

Paterson, M. 2009. Haptic geographies: Ethnography, haptic knowledges and sensuous dispositions. *Progress in Human Geography,* 33(6), 766–78.

Paterson, M. 2010. More-than-visual approaches to architecture. Vision, touch, technique. *Social & Cultural Geography,* 12(3), 263–28.

Pearce, M.W. 2008. Framing the days: Place and narrative in cartography. *Cartography and Geographic Information Science*, 35(1), 17–32.

Rodaway, P. 1994. *Sensuous Geographies: Body Sense and Place.* London: Routledge.

Rose, G. 2003. On the need to ask how, exactly, is geography "visual"? *Antipode*, 35(2), 212–21.

Saville, S. 2008. Playing with fear: parkour and the mobility of emotion. *Social & Cultural Geography,* 9, 891–914.

Serres, M. 2008. *The Five Senses: A Philosophy of Mingled Bodies.* London: Continuum.

Sobchack, V. 2004. *Carnal Thoughts: Embodiment and Moving Image Culture.* London: California University Press.

Sparkes, A. 2009. Ethnography and the senses: challenges and possibilities. *Qualitative Research in Sport and Exercise,* 1(1), 21–35.

Spinney, J. 2006. A place of sense: a kinaesthetic ethnography of cyclists on Mont Ventoux. *Environment and Planning D: Society and Space,* 24(5), 709–32.

Stoller, P. 1997. *Sensuous Scholarship.* Philadelphia, PA: University of Pennsylvania Press.

Thrift, N. 2007. *Non-representational Theory: Space, Politics, Affect.* London: Routledge.

Winnicott, D.W. 1971. *Playing and Reality.* Harmondsworth, England: Penguin.

Wylie, J. 2002. An essay on ascending Glastonbury Tor. *Geoforum,* 33(4), 441–54.

Wylie, J. 2006. Smoothlands: fragments/landscapes/fragments. *Cultural Geographies,* 13(3), 458–65.

Chapter 1

Negotiating Therapeutic Touch: Encountering Massage Through the 'Mixed Bodies' of Michel Serres

Jennifer Lea

Introduction

In this chapter I consider how touch is used towards 'therapeutic' ends in the context of massage in the UK. Massage is a form of 'interactive embodied work' (McDowell 2009) or 'body work' (Wolkowitz 2002); wider categories of service employment where work is done on the body, towards such ends as health, beauty or well-being. The category of body work includes such occupations as nursing, care work, beauty therapy, hairdressing and massage. These forms of work are becoming significant in terms of the numbers of workers employed, as well as the cultural and social significance of such work for the production of modified bodies and selves (see McDowell 2009, chapter 2). Existing studies of forms of body work have predominantly attended to the gendered, sexualised and classed division of labour involved, in which naturalised assumptions about embodied attributes and gendered performances have meant that the labour force consists largely of female workers. McDowell and Wolkowitz have critiqued the way in which the emotional and physical effort involved in the work has been obscured by such assumptions about gendered identities and work, and their research has significantly advanced understandings of the social relations within which these body workers operate, and which the work serves to reproduce.

Despite the central position which embodiment has come to occupy in these studies, this focus on embodied structures has tended to obscure the body itself. While Wolkowitz (2002, 497) defines body work as involving 'intimate, messy contact with the (frequently supine or naked) body, it's orifices or products through touch or close proximity', the consequences of this intimacy, messiness and close proximity have not been fully evaluated. The focus on structures of embodiment has tended to frame the body as a straightforward and bounded object that can be read through its social and cultural characteristics. Thinking about the body in this way has meant that a more phenomenological concern with the experiences, movements and the corporeal nature of the transactions taking place in body work remains under-examined. The effects and consequences of body work have largely been conceptualised as being either in the emotional register (Wolkowitz 2002,

500), or on the outside of the body (e.g., inappropriate touch mapped across the surfaces of the body), leaving the relation between these feelings and the actual corporeality within which they are located and across which they are distributed, unexamined. The specificities of the body-body relations are absent; the particular types of touching, manipulation and forms of relationship have not been specified, nor their effects explored. Touch, for instance, has generally been assumed to be a constant not only within one form of body work (e.g., massage), but also between different forms of body work (e.g., massage and nursing).

This chapter seeks to advance understanding of these experiential, felt, fleshy registers of the body, doing so by drawing on in-depth interviews conducted with a variety of massage practitioners based in major UK cities (London, Bristol, Glasgow and Edinburgh). In these interviews, the respondents were asked specifically to reflect on their use of touch in massage. This interview data allows several questions to be examined: how does the massage practitioner use touch to 'diagnose' and 'treat' the ailments of clients? How are the bodily effects of touch experienced and understood in the context of massage? In what ways do massage practitioners interpret the reactions of clients to their touch? What are the interrelations between touch and intimacy? What is intimate touch? What are the consequences of intimate touch? The chapter draws on the work of Michel Serres to signpost its analysis towards these fleshy 'registers' (or parts of the body). Serres, in his book *The Five Senses*, describes the body as an open, 'mixed', entity that is related to other bodies and the world through the action of the senses. He uses topological metaphors to explain the folding together of body and world that occurs as sensory excitation breaches the boundary of the skin. His work on touch is useful because it has resonances with the practitioner's accounts of the action of touch beyond the surface of the recipient's skin, and it also provides one way forward for thinking about the fleshy, corporeal effects of body work. The chapter begins by giving a background to massage and the methods used to collect this data, before outlining the work of Serres on touch, and then examining the use of touch by the massage practitioners.

Introducing Massage

Simply put, massage involves the manipulation of the soft tissues of the body through touch. The effects of touch also manipulate the bodily experiences of the client. During a massage a practitioner might touch, move, manipulate and stretch almost all the parts of the client's body. While moving around the body, or dwelling on particular points, the practitioner can use their fingers, hands, elbows and/or feet to touch their client. Dwelling on particular points, or rubbing large areas might be done directly onto the skin or through clothes. Objects and substances such as hot stones or aromatherapy oil might also mediate the touch. Different types of massage use different types of touch, and often practitioners are trained in a variety of kinds of massage. Massage is generally understood as belonging within the wider category of 'complementary and alternative medicines', and as such has been associated with the development of contemporary cultures of well-being (Sointu 2006). Here, the body has been positioned as an arena within which 'personal quests for health and well-being' (Serres 2008, 272) might be pursued.

With this significance being accorded to the body, massage as a form of body work has grown in importance in the UK, such that a wide (and proliferating) range of types of massage are now offered in many different settings. While the broader research within which this chapter is situated looked at massage in a variety of contexts (training, a spiritual retreat, and a 'healing area' at a British music festival, cf. Lea 2009a, b), here the focus is on massage as delivered within 'natural health centres'. These centres provide rooms that massage and other complementary and alternative medicines practitioners can rent in order to treat their clients. These clinics provide an organisational structure within which the practitioner works; a receptionist manages the client's bookings, arrival, payment and departure. Sometimes there are a variety of herbal medicines, creams, tinctures, remedies, beauty products and books for sale (cf. Doel and Segrott 2004). Clients wait in a central waiting area and are then taken to the room where the practitioner will deliver the massage, where there is often a massage couch or futon, and a small consultation area. Natural health centres most often have multiple treatment rooms so a number of practitioners can work at the same time.

The practitioners interviewed for this research were recruited through natural health centres, and consequently they practice a range of different types of massage, involving a variety of ways of touching clients. Rather than attempting to systematise the use of touch according to the type of massage, the chapter offers some opening reflections on the different ways that practitioners use and understand touch. All the interviewees had upwards of four years experiencing of working full-time as a massage practitioner, thus developing their 'haptic skills' (Hetherington 2003, 1934. See also Thrift 2007, 103). This degree of experience meant that the interviewees found it hard to give a phenomenological account of the 'doing' of massage (see Lea 2009a), instead finding it easier to explain their understandings of touch. As a result, the focus of this chapter is on the ways that the interviewees make sense of the action of their touch on the client. While the chapter does not utilise client accounts of massage, these practitioner interviews offer a significant starting point from which to consider massage.

Understanding the Body

In a recent paper on the geographies of touch Dixon and Straughan (2010, 450) indicate that, while it is 'an integral, everyday component of modern-day Geography', touch has received little sustained critical attention. The predominantly visual nature of geography has, with some exceptions, obscured the other senses. However, recent 'non-essentialist' geographies of the body have, they note, advanced understandings of the nature and action of touch. These theorisations refuse to understand "the body" as a 'primarily autonomous or self contained system' (Dixon and Straughan 2010, 450), suggesting that system-based accounts problematically place structures prior to the body (Thrift and Pile 1995, Dewsbury 2000). Non-essentialist accounts suggest that the body is an open entity in which the 'interiority' of the body (the 'psyche' and 'the meat, flesh and bones of the soma') is connected with the 'exterior' world (Dixon and Straughan 2010, 450). Geographers have drawn on theorists such as Deleuze and Guattari (Paterson 2005) and Irigaray (Colls 2011) to offer accounts of the formation of the body, mediated through touch.

Serres offers another way to understand how this body-world conjugation is 'made' through the action of touch. In his book, *The Five Senses* (2008), he argues that the significance of the senses and the body to our subjecthood has been obscured, because language has become primary in Western cultures. To bring the senses and the body back to the fore Serres attends in detail to the sensory relation between body and world. In all of Serres's writing, the *relation* is of primary importance, being seen to 'spawn objects, beings and acts' rather than 'vice versa' (Serres and Latour 1995, 107). It is the way that bodies relate through the senses that allows Serres to theorise both *the kinds of bodies that are made* by sensory relations and *the action of the senses in making* those bodies. Rather than apprehending bodies as systematic predictable objects, Serres argues that they must be approached via their mixing with other objects, bodies and contexts. This mixing is ongoing (Serres 2008, 80) but is often obscured by the 'violence' of the linguistic treatment of the body as a bounded object (Serres and Latour 1995, 132). For Serres this mixing is not partial but rather radically refigures what is commonly understood as "a body". He writes that 'two mingled bodies do not form a separate subject and object' (2008, 26) but instead combine.

The relation between body and body (or body and world) implied here is one that is not contained at the surface of the body but rather extends into the interior. The senses provide a lens through which this process of mixing might be viewed. Connor (2008, 3) notes that for Serres the senses *are* the mixing of the body: 'the principal means whereby the body mingles with the world and with itself, overflows its borders'. The senses are understood to disrupt the boundaries of the body and the structures of the body: working to mix body and world and interior and exterior. However, Serres notes that the action of the senses has been 'black boxed': while the sensory inputs into and outputs out of the body have been surveyed (the intersection between the body and the physical world as a space of sensory variability, and the bodily effects which we work into a meaningful experience), the work of the senses in making and remaking the body has not been fully scrutinised. Serres thus pays close attention to the action of the senses, directing the reader to think of sensory excitations moving from 'outside' to 'inside' via a mechanism of 'kneading'. Connor (2002, 5) gives a helpful description of this process: 'in kneading, one repeatedly folds the outer skin of the substance inwards, until it is as it were crammed with surface tension, full of its outside'. Using the image of a loaf of bread being kneaded, the action of the senses is seen to fold the 'exterior' into the 'interior', thus rearranging the organisation of the body: the dough moves in an unpredictable fashion: with each fold proximate points are dispersed and previously distant points brought together.

The mixing of body and body, or body and world brings unpredictability into what has generally been understood to be a relatively predictable and stable system, and underlines the tension that exists between thinking about the mixed body and the body as contained by the skin. While Serres goes so far as to suggest that the 'construction of the body proper' is a 'fiction' (2008, 62), this idea of the body remains the predominant one inhabiting Western cultural imaginations. Serres recognises this tension in his discussion of touch and the skin. He notes that the skin has dual and contradictory roles: on the one hand, as already described, the skin is permeable and porous, acting as a heightened surface for relating-by-kneading and offering access to the fleshy substance of the body. The skin is the first meeting point of body and world: 'in it the world and the body intersect and caress each other' (Serres 2008, 80). At the same time, the skin holds the body together, delineating it as a bounded systematic wholeness, and holding organs, blood and corporeal fabric together. Serres (2008, 55) writes that the skin 'imprisons' the body: in this way 'agonist of a kneaded world, for the skin is what holds individual lives separate and aloof' (Connor 2004, 5).

What the kneaded, mixed body of Serres offers to geographers is another way to think about touch which emphasises the way that the body is made and remade in relation to it's context. Serres complicates the distinction commonly made between the interior and exterior of the body (see Paterson 2009), and emphasises the connections between the different elements of the body's interiority (the fleshy, the hormonal system, the emotional for instance). The image of 'kneading' offers a way of imagining the sensory relation between body and world, and how touch plays a part in reworking bodies (beyond the skin). In foregrounding the tensions inherent in the body – between ordered system and the disordering effect of sensation, or between the contained and the excessive body – Serres offers a clear reminder that the body is contradictory and multiple (Dewsbury 2000).

Mixing and kneading are two ideas that are particularly instructive in the analysis of massage, directing attention towards two key questions. Firstly, how practitioners use touch to diagnose and treat 'problems' within the client. Secondly, how practitioners and clients manage the unpredictable effects of touch in treating these 'problems'. Of particular interest here is how boundaries of (in)appropriate touch are marked not only on the surface of the body, but also become manifest through this 'interior' action of touch.

Understanding Massage

Without exception, the practitioners I interviewed described massage in broadly therapeutic terms. While they all understood touch to have a role in solving particular bodily 'problems', their accounts of the causes of and solutions to these 'problems' differed. All of them began their massages with the taking of a case history; part of the 'constant *performance* of professional legitimacy' (Oerton 2004, 545 original emphasis; see also Doel and Segrott 2003a, 2003b, 2004) demanded by the contradictory position of massage and other forms of complimentary and alternative medicine within the wider landscapes of healthcare provision in the UK. They asked clients about work, why they had come for a massage and any physical or emotional issues that they thought the practitioner might be able to help with. For many of the practitioners this verbal questioning did not offer them enough information so they turned to the body of the client to give them more details about the 'problems' that the client had. Jane, a practitioner based in Edinburgh, told me that she starts looking for 'problems' in the client as soon as she calls the client from the waiting room into the treatment room,

> they have to walk up the corridor because they'll be sitting down so, you can see how they're seated before you go through, see how they are kind of facially, see if they've got pain or are happy, or so kind of see a bit of their character. Um when you're sitting down it's very easy to see maybe if they're lopsided, how they sit while they're at work because, if they're kind of quite slumped. (Jane 23/01/2008)

Jane looks for the way that 'problems' affect the posture and gait of the moving body, as well as being shown in facial expressions. After the case history, she then begins to use touch in an exploratory, diagnostic way. Jane said she found it 'good to spend a bit of time when you're not actually massaging' to get a 'feel for what is happening and for them to give you feedback – so what does it feel like to them? Does it feel sharp, does it feel tingly?' Jane uses a combination of the body's account and the client's verbal account to get an idea of the 'problems' that she might address through massage. Jon, a Thai massage, therapeutic massage and Indian head massage practitioner based in Glasgow found less value in the client's verbal account, preferring to use touch as his primary diagnostic tool. He told me about the kinds of touch he used to diagnose 'problems' within the client:

> [I] use my palms quite a lot 'cause that's a nice broad touch that is much more acceptable to lots of people so when I first start … I'll put one hand between their shoulders, one hand at the base of their spine and just hold it there. And it's just them getting used to me, me getting used to them and through doing that I can actually sense where there's areas of tension. So then after they've got used to that then I'll work my hands gently, just gently pressing up and down the spine … so it gives me a guide of where the stresses and strains are but also it's relaxing for them to get that pressure up and down the spine. Erm with the Thai [yoga massage] I start off by doing … by going round the whole body and just literally holding on to each of the joints so it would be hold on to the ankles, knee, hip, erm solar plexus, shoulder, elbow, wrist… (Jon 28/02/2008)

This kind of diagnostic touch indicates a belief that the client's 'problems' are located in, and treatable through, their body. The practitioners told me that the 'problems' are not just physical, but might also be mental or emotional. The physical body is believed to be an archive of past practices and experiences; physical effects might result from physical stresses or strains, or the effects of stored emotions or feelings (for instance, tight or knotty muscles are understood to be a permanent record of the physical and emotional history of the body). In Serresean terms, 'problems' arise from the past experiences that have been kneaded into the body and, as a result, constitute its present corporeality. Diagnostic touch is static or gently moving, aiming to explore rather than change the body. For instance, Jon holds joints like the ankles and knees to gain familiarity with the contours of the body's surfaces and gently palpates to access the record of past traumas that are being held in the interior of the body.

I asked the practitioners how they felt a 'problem' and they described differences in the body's tissues. Jon described how muscles can feel variously like different kinds of rubber: 'so you've got for example, a dog's rubber bone – really solid rubber...Or you've got rubber like putty, so some muscles are like putty and they are much more malleable'. Annie, a practitioner who performs massage in pregnancy and Swedish massage in Glasgow, used a couple of clients to give examples of the feeling of problematic areas to her. She told me that one client had 'little knots all the way up her spine, just in the mid-thoracic ... they're just like little counters under skin. I kind of feel like little round buttony type tension' (Annie 5/6/2008). Another had shoulder issues and had 'gristly muscle tissue ... so when you run your hand over it, you feel its like you feel the little grains or something moving under your thumb, under your finger'. Here, Annie is equating the feel of the body with the experiences it has had; the gritty or fibrous muscles indicate ingrained trauma and repetitive strains. The physical fibres of the body are linked with emotional responses. The practitioners associate what they touch with a visual checking of the client's reactions: I was told that flinching or ticklishness might signpost physical pain or areas of emotional distress, showing how physical and emotional trauma are folded into the body hybridising the flesh. These bodily reactions indicated to the practitioners the existence of a 'problem' held in the body, which required reworking through their touch.

At work in the way that the practitioners locate physical and emotional 'problems' in the body is a sense that touch affords privileged access to the interior of the client's body. The implication here is that the body can tell its own story, and some of the 'problems' that are stored in the body are potentially inaccessible to the cognitive thought or experience of the client. Jon and Kay, holistic massage practitioners from Bristol, expressed some kind of distrust of the verbal account when they told me that the client's body was able to give a fuller account of its 'problems' than the client was able to in the case history. Jon told me that he uses diagnostic touch to assess whether the account given to him in the verbal consultation matches the account given to him by the body during the preliminary touching:

Jon: you do the consultation, so people say they come along with a problem that they think they've got, but there are always underlying problems so I use my hands ... I use my hands and then you can actually say: 'well, you've got this there, and you've got this there' and they go 'I forgot about that' ... the issue that people come along with is rarely the underlying general overall issue; its like a high note. It's the loudest thing in their head.

Jen: kind of present at that particular moment?

Jon: exactly, but all the other stuff which is a lot lower and hums a lot more is a lot deeper, and it's a lot deeper sort of psychologically, emotionally, physically...

Kay told a similar story based on her own experiences in which her own 'problems' were detected by a massage practitioner:

I went for a massage, and actually, that was the beginning of the change, because I had to shut up. My body told the person exactly what was going on, and that person had to treat the symptoms in my body ... we all know how wordy we can be, and how easy it is to tell untruths, without even wanting to. I mean I wanted to speak the truth, but I didn't know really what the truth was (Kay 6/03/2008).

A similar notion of a bodily truth that is inaccessible to the client's cognitive awareness is present in Jane's account, who told me that her touch might direct her to work on a different area of the body than the location of the pain reported by the client in the verbal case history:

'so they might want their back done when you start, and then you might end up working on their leg because it's that area that is really tight' (Jane 23/01/2008).

These quotes demonstrate how some practitioners privilege the senses in a similar way to Serres. As already noted the central argument in *The Five Senses* is that for too long language has separated the western subject from the senses and the body. Language, Serres argues, has replaced experience, enacting a kind of 'violence' on the sensory body (Serres and Latour 1995, 132). It is only by leaving language behind that the flesh is 'freed' and it is possible to 'get back' to the senses (Serres 2008). The practitioners understanding that the sensory realm affords them privileged access to the client's problems, however, requires more inspection, particularly in the context of Wolkowitz's (2002, 503) warning that in analyses of body work 'we should be wary of assuming that practitioners' and patients', clients' or customer's interests necessarily coincide'. While touch can be seen as a 'model for sympathy, of literally *feeling-with*' (Wyschogrod 1981, original emphasis, cited in Paterson 2007, 147) allowing the 'transfer of sympathy and empathy between individuals' (Edwards 1998, cited in Paterson 2007, 153), this empathy can be in tension with the excessive and unpredictable action of touch as well as the authority of the practitioner to access and fix the 'problems' that lie deep within the client. The interviews offer an opportunity to begin to consider how this tension manifests in practice.

The practitioners used touch as a tool to access and release the 'problems' that they see to be located in the interior of the body. While this idea of 'release' was common, the practitioners used different terms to describe it. For instance Tina, a practitioner from Glasgow who practiced deep tissue, aromatherapy and Indian head massage, told me about the effects of touch on a physical level:

> for example there is a point in the glutes which often is quite sore for pregnant women ... If you work those points they are very sore and tender ... but what you will hopefully find is that pain will ease off and it will literally ease off in a couple of minutes ... you can feel like the grittiness and knots of where things are. And when I am working with acupressure points what I will do is tend to work those points at least three times in rotation so by the third time it should have eased off. (Tina 31/01/2008)

Annie described a similar physical effect of repeated touch on the 'knots' or tensions she described, telling me that:

> some of them even 'pop' so that when you press them they just go 'crunch, crunch, crunch', or 'pop' and that's a great feeling. When it happens you know you've achieved something and the person goes 'ah, perfect! (Annie 5/06/2008)

These accounts describe the action of touch on the physical fibres of the client's body to release the physical tensions and stresses stored in the body, for example the negative effects of repetitive movements or bad posture. Other practitioners, additionally, described the release in emotional terms, telling me that touch allowed them to access and release past emotions and experiences stored in the physical fabric of the body's interior. Jon used an experience with one client (who had a 'niggling' problem in the area between the shoulder and neck) to describe this link:

> [I] found this ... nodule of tension and I worked it, worked it, worked it, and then it popped so it disappeared. But at that precise moment that it popped, shooting up into her brain was this memory of a car accident she'd had. So what it was was whiplash from this car accident and she'd stored that tension in her body ... the stresses and strains are built up in the body and it's linked with the mind ... She'd obviously got whiplash and held herself and it was held. It was like a time capsule held in her mind and it was perpetuating that muscle tension there. (Jon 28/02/2008)

Similarly, Zoe, who practices Thai massage and zero balancing in London, used her own experiences of having a massage in Thailand to talk about 'emotional' release. She described how the massage practitioner:

gets up to my solar plexus, and he's moving and he's moving and he's palming and he's thumbing and he's moving and he's rocking, and its not particularly dynamic stretches. Suddenly there's this emotion that comes up and I'm crying (and you know, the Thai's are like 'oh whatever!' and they carry on and not like 'oh, do you want to talk about it?'). And for me it was like 'ah, I've hit upon something'. I've always known physically that's where I hold – suddenly I've encountered this again through bodywork. Maybe I could use bodywork to deepen my understanding of what is it that is held in here? (Zoe 14/05/2008)

This suggestion that touch can precipitate the cathartic release of deeply held emotions can be seen in the accounts of a number of other practitioners interviewed, as well as in analyses of other forms of body work (e.g., Paterson's 2005 writing on Reiki). The model of kneading and mixing outlined by Serres offers one way to understand how this release might happen. The kneading action of the touch allows the practitioner access to the interior of the body where the 'problems' are seen to be located. The disordering effect of touch rearranges the bodily archive such that past experiences might be 'released' from their location and eventually even released from the body. The disorder effected in a system (e.g., the body) by kneading is described well by Connor (2002, 2):

> Serres imagines trying to map or model the involutions of the dough as it is molded, perhaps by making a mark and plotting its changes of position … the trajectory of this point relative to other points in the dough would very quickly become undetermined, irrational, as seemingly random as the flight of a fly.

For massage practitioners I interviewed, touch enables them to work on both the physical body and the emotional life of the client. For Jon, this idea of releasing 'problems' from a client's body became a kind of therapeutic 'imperative'. He told me about his belief that bodies are not able to sustainably carry about 'problems' and that unresolved negative experiences damage the body: 'a lot of a lot of people have mental, emotional problems which they don't deal with, so they store them and they tend to store emotional problems as physical tension'. Jon understood his touch to intervene in the fabric of the body to make 'people to remember things … the deeper down stuff … and through using touch I help people get back in touch with it, and hopefully release it so it's no longer an issue'. His understanding of the effectiveness of touch in dealing with emotional 'problems' was such that he, somewhat contentiously, positioned touch as an alternative to talking cures such as counselling:

> they don't need to actually talk about these problems because they've stored these problems in their body and what I'm doing is I'm using my touch to actually help those problems by releasing the tension in their body. (Jon 28/02/2008)

The strength of Jon's belief in the action of touch meant that he found client's resistance to his touch problematic. He described a number of scenarios where the strength of the client's bodily boundaries was resistant to the mixing and kneading of his touch, maintaining the separation between client and practitioner and not allowing him access to the 'problematic' interior. For instance, he told me about one client whose body was not receptive to his touch:

> I had one particular client who came to me a year ago and she was so locked in her mind that she'd had real emotional problems, had a depression, was a teacher in a high powered expensive school and she was coping but her life was in shreds but she was still coping and her counsellor said you've got to go along for a massage so she came along for an Indian head massage and initially I could only literally touch her head and shoulders and that was about it. After about five treatments of Indian head massage she actually said: 'I think I might actually relax enough to have a back, neck and shoulders massage, lie down take my clothes off for you to actually massage oil in to my back'. After ten treatments she had a full body massage. (Jon 28/02/2008)

In this case the client persisted with the massages even though the touch was uncomfortable, and Jon extended the model of touch as sympathy and empathy, advancing slowly to suit the client. Jon described further how other clients were able to accept touch up to a certain point but once 'they've actually got to a level whereby they're just about to let go ... they don't want to let go'. He also explained that other clients deny that the touch is affecting them because otherwise the effects would threaten the coherence of their body and self and would be too difficult for them to cope with:

> they say [the massage] has not affected them, whereas you can see that it has affected them, but they won't allow it to have affected them ... Because they can't be seen to let go because there's so much stored inside, that it's like opening ... Pandora's box and too much will be let go. (Jon 28/02/2008)

This denial of the effects of his touch was a source of frustration for Jon, partly because of the strength of his belief in the necessity of releasing 'problems' from the body. He described some clients as *unwilling*, rather than *unable*, to yield to the action of touch and suggested that this was indicative of the existence of deep-seated problems within the client:

> some people have their body's armour and you can feel it...there are some people that won't let go and you can work on someone and they've got so many stresses and strains and it is almost like armour...their muscles are armour plating and some people will let you get through and relax and some people won't. (Jon 28/02/2008)

Other practitioners similarly commented on the reluctance of clients to 'open up' and allow them access to their 'problems'. Jane, for example, told me about some of her clients who had chronic or long term problems and were unwilling to let go of their 'problems' because they were unwilling risk that the touch might make them feel worse before they felt better:

> I find it quite interesting where people have chronic things, like what they want to hold onto, so they don't actually have to go through any more if that makes sense? … So if they're really tired all the time they don't want to feel more tired … we kind of hold onto things until we can deal with them, but sometimes there's not really a point where we can deal with them, so how much is everybody actually holding onto all the time? Erm, you know, probably for years, because they just go, 'right, [I] can't deal with that' (Jane 23/01/2008)

The theme running through both of these accounts is that some clients are resistant to the action of touch because they see their 'problems' to be integrally constitutive of the fabric of their bodies. The clients are protective of the experiences that make them up, so that releasing their 'problems' is too much of a radical disruption to the self for them to contemplate. The clients are seen to utilise the skin's properties as a boundary to resist the mixing and disordering effects of the practitioner's touch. While the idea of releasing 'problems' can be seen to be central to the accounts of most of the practitioners, they also described how they had to exercise caution in their use of touch to avoid the kind of extreme cathartic release described by Zoe. Jon explained this in terms of the organisational structures within which he worked and the practicalities of earning a living as a massage practitioner:

> With people crying I tend to, I tend to be quiet, I tend to let them get on with it because people tend to cry, say I'm sorry and then maybe explain the reason why they're crying. I don't want to put too much on to that, I'm not there to be their parent or anything. I'm not there to sort of say 'there, there, there' or even say 'what's the problem?' Because then you're opening a can of worms and then you can be there for another half an hour and if they're crying at the end of the treatment and you've got … another client outside and you know mentally that this is a business, you've got to keep your business head. (Jon 28/02/2008)

Jon has to manage his expectations about his ability to work strongly with the clients, balancing his short-term responsibilities to his other clients booked into later time-slots, and longer-term responsibilities to himself and his family to earn a living and build a sustainable career without experiencing 'burn-out' (see also Oerton and Phoenix 2001, 393). Tina offered a different account stating that it was not her role to provide emotional support because she is not trained in this way. She said:

> I don't tend to emotionally support the clients because I am not a trained counsellor, and I don't want to put myself in that position, so what I'll do is I

tend to listen and say 'right well, yeah I understand that' and if there's a natural remedy I can recommend. I try to offer a practical angle. (Tina 31/01/2008)

Given that there is a 'sharp cutting-off point' between touch that is acceptable and unacceptable (Paterson 2005, 161), it was not always possible for the practitioners to avoid cathartic release. While the surfaces of the body are culturally divided into those that are acceptable and unacceptable for the practitioner to touch, the effects of that touch are unpredictable. The practitioner cannot know in advance for sure whether their touch will cross their client's thresholds or limits of acceptability. Furthermore, because the practitioners are using touch to discover 'problems' that the client is not necessarily aware of, they cannot warn the practitioner about those limits. Gentle touch might result in a particular strong reaction without the client or the practitioner being aware of this in advance. As such the practitioners told me about strategies they used with clients to mediate the effects of their touch. Tina, for instance, described using a different kind of touch to help calm a client who had suffered a panic attack during the massage. She told me:

One of my clients had emotional stuff going on, and we discussed it – what she felt. And then we did the massage, and about half-way into the massage as she relaxed, she had a panic attack ... 'cause she'd been to me before, she was able to say that so I just covered her up with a sheet and then did holding points. So it's like the back of the neck and it's just very calming and then just get her to rest. So once she'd calmed down ... [I] just gave her the option did she want to continue with any more of the treatment. And she actually was happy to, and so what I did was take it away from where it felt intimate, so did the hands, arms. (Tina 31/01/2008)

Jon saw value in shortening the treatment and talking to some clients to enable them to make sense of what had happened, remarking that 'obviously it's all part of the treatment, so what's in there needs to come out'. Both of these techniques can be see as strategies to bring order back to a body that has been disordered by the effects of touch. For example, Jon talks to the client so that she can make sense of the sensations, while Tina held the outside of the body, utilising static touch to bring the boundaries of the body to the forefront of the client's awareness. Tina and Jon play a part in reordering the body into a coherent system that the client is able to make sense of, thus reinstating their sense of corporeal wholeness.

Having indicated some of the strategies used to manage touch, it is important to note that not all the practitioners saw the release of 'problems' as their primary goal. As such it is important to take note of some of the other ways in which they explained touch. One way was the use of touch to increase the client's self-awareness, as described by Jane:

it's allowing you to let the patient tune into themselves, and you're just tuning into something as well, but you're not having to be invasive and you're not causing them pain. So I think as a therapist I don't like causing them pain as such … I'm kind of at odds with trying to do something that is any deeper or more painful, so I'd like to find gentler routes to it. (Jane 23/01/2008)

In saying that her touch allows her clients to 'tune into themselves', Jane suggests that the sensory experience of being touched draws the client's attention to their bodies. Zoe expressed a similar belief, explaining that:

to me it's the movement and it's the attitude and the approach of the practitioner … your desire is to be completely present, completely open hearted, completely unjudgemental – rather than like 'ooh, you should be a bit looser there so we'll try and stretch you', which some people want that … Actually I think that the real healing comes when we're not trying to do anything; we're just presenting … [for example] this weird sensation in this solar plexus – we're not trying to loosen it by pressing it a lot, or we're just going 'oh look, there it is, there it is'. Then I see that the baton is handed over to [the client]: 'what, how do I react to this, what do I do with it?' Do I go 'goddam its that feeling again in my solar plexus' or do I just sit with it and let it be? … That that's the real healing depth – it's almost like it's a way of meeting and accepting your own body. (Zoe 14/05/2008)

Jane's aim was to allow the client to 'get to a place where [they] could listen to what's going on' in their bodies. Rather than understanding touch to give her privileged access to the client's problems, Jane sees touch as enabling the client themselves to 'return' to their bodies. The kneading action of touch is seen to allow the practitioner to access the interior of the client, with the aim of bringing the 'problems' to the cognitive awareness of the client. Zoe extends this idea suggesting that the opportunity for real healing lies in the use of touch to bring awareness and acceptance to the self, without making judgements about the 'problems' that lie within the client, or bringing an imperative to change the body. Zoe's understanding that the sensory experience of touching and being touched needs to be detached from the process of judgement making bears some resemblance to Serres's argument that language has to be left behind in order for the Western subject to 'get back' to the sensory realm. For Jane and Zoe, their touch offers the client the possibility of knowing, understanding and contesting 'problems' in their body, placing the responsibility for the management of touch in the hands of the client as well as the practitioner. The most obvious line of analysis here would be one that suggests Jon imposes his therapeutic will on his clients in a potentially problematic way, while Jane and Zoe set up more symmetrical power relations. However, an alternative reading might be that Jane and Zoe are irresponsibly opening up the body as a 'problem' and passing the responsibility for this to the client who may be ill equipped to negotiate this. As Doel and Segrott (2003a, 751) note in the context of complimentary and alternative medicine more generally, 'with empowerment comes a perilous responsibility', and these very bodily ideas of empowerment and responsibility within the body work relation certainly require further analysis.

Conclusion

Tracing the complex connections between the body, flesh, skin, touch and the emotions has underlined the value in attending to the experiential and felt qualities of 'therapeutic' forms of touch. The practitioner's accounts of touch reflected the multiple and diverse nature of touch, underlining the need for attention to be paid to the 'detailed somatic processes or haptic knowledges' (Paterson 2009, 781) that are integral to touch. In continually returning to the contested and negotiated nature of touch in the massage relation, the chapter underlines the necessity of attending to the experiential and felt qualities of therapeutic touch, and retracing the connections between touch and the therapeutic, across a range of body work contexts. Conceptualising and understanding these body-body relationships enacted through touch is crucial in further developing understandings of body work, where 'how bodies connect (or do not) is a key issue' (McDowell 2009, 225).

The ideas of kneading and mixing have value in understanding these connections between bodies. Mixing offers one way to understand the action of touch across and between separate bodies, reflecting recent geographical research that troubles the notion of touch being clearly delineated within an individual body (Paterson 2009, 779). Kneading allows the action of touch to extend across a range of bodily 'registers'; as well as the physical action, it can affect the emotional, or the energetic for instance (cf. Lea 2009a). These concepts highlight two aspects of body work that briefly arose in the chapter but deserve more sustained attention. Firstly, they extend the idea of 'intimacy' so that intimate touch means not only touching the 'outside' of the body in areas culturally or socially understood as intimate (e.g., the stomach, the face, the ears), but also in affecting intimate and deeply held sensations, feelings and experiences. The consequences of this kind of intimate touch for both client and practitioner need further examination. Secondly, the relationship between touch and language in the context of massage (and other forms of body work) needs further analysis. While Serres argues strongly that language has obscured the sensory realm, a focus on client experiences might suggest a more complex relation between language and touch, in which the sensory intensity of massage might take the individual away from language (giving them a rest from the overly busy narrative of their minds), but for others the sensory might deliver them to language (in which they might attach judgement to the way that touch exposes their bodies to them; cf. Lea 2009b). Both of these require examination of client experiences to 'flesh out' the experiential realm of touching and being touched.

Acknowledgements

The research that this chapter was based on was funded by the Department of Geographical and Earth Sciences, University of Glasgow. Thanks are due to Chris Philo and Hayden Lorimer who offered helpful comments on an earlier version of this chapter, and to the editors for their comprehensive comments and patience.

References

Colls, R. 2011. BodiesTouchingBodies: Jenny Saville's over-life sized paintings and the 'morpho-logics' of fat, female bodies. *Gender, Place and Culture,* DOI: 10.1080/0966369X.2011.573143.

Connor, S. 2002. *Michel Serres's Milieux.* Mimeo, <www.bbk.ac.uk/english/skc/milieux/>.

Connor, S. 2004. *Topologies: Michel Serres and the Shapes of Thought.* Mimeo, <www.bbk.ac.uk/english/skc/topologies/>.

Connor, S. 2008. Introduction, in *The Five Senses: A Philosophy of Mingled Bodies,* by M. Serres. London: Continuum, 1–16.

Dewsbury, J.D. 2000. Performativity and the event: enacting a philosophy of difference. *Environment and Planning D: Society and Space,* 18, 473–96.

Dixon, D. and Straughan, E. 2010. Geographies of touch/touched by geography. *Geography Compass,* 4/5, 449–59.

Doel, M. and Segrott, J. 2003a. Beyond belief? Consumer culture, complementary medicine, and the dis-ease of everyday life. *Environment and Planning D: Society and Space,* 21, 739–59.

Doel, M. and Segrott, J. 2003b. Self, health, and gender: complementary and alternative medicine in the British mass media. *Gender, Place and Culture,* 10, 131–44.

Doel, M. and Segrott, J. 2004. Materializing complementary and alternative medicine: aromatherapy, chiropractic, and Chinese herbal medicine in the UK. *Geoforum,* 35, 727–38.

Hetherington, K. 2003. Spatial textures: Making place through touch. *Environment and Planning A,* 35(11), 1933–44.

Lea, J. 2009a. Becoming skilled: the cultural and corporeal geographies of teaching and learning Thai Yoga massage. *Geoforum,* 40, 465–74.

Lea, J. 2009b. Liberation or imitation: Understanding Iyengar Yoga as a practice of the self. *Body and Society,* 15(3), 71–92.

McDowell, L. 2009. *Working Bodies: Interactive Service Employment and Workplace Identities.* Oxford: Wiley-Blackwell.

Oerton, S. 2004. Bodywork boundaries: power, politics and professionalism in therapeutic massage. *Gender, Work and Organization,* 11(5), 544–65.

Oerton, S. and Phoenix, J. 2001. Sex/bodywork: discourses and practices. *Sexualities,* 4(4), 387–412.

Paterson, M. 2005. Affecting touch: towards a felt phenomenology of therapeutic touch, in *Emotional Geographies,* edited by J. Davidson, L. Bondi and M. Smith. Aldershot, England: Ashgate, 161–76.

Paterson, M. 2007. *The Senses of Touch: Haptics, Affects and Technologies.* Oxford: Berg.

Paterson, M. 2009. Haptic geographies: ethnography, haptic knowledges and sensuous dispositions. *Progress in Human Geography,* 33(6), 766–188.

Serres, M. 2008. *The Five Senses: A Philosophy of Mingled Bodies.* London: Continuum.

Serres, M. and Latour, B. 1995 *Conversations on Science, Culture and Time.* Ann Arbor: University of Michigan Press.

Sointu, E. 2006. Recognition and the creation of well-being. *Sociology,* 40(3), 493–510.

Thrift, N. and Pile, S.1995. *Mapping the Subject: Geographies of Cultural Transformation.* London: Routledge.

Thrift, N. 2007. *Non-Representational Theory: Space, Politics, Affect.* London, Routledge.

Wolkowitz, C. 2002. The social relations of body work. *Work, Employment and Society,* 16(3), 497–510.

Chapter 2
Touching the Beach

Pau Obrador

A Beach without Bodies

Tourism is becoming sensually more diverse and this is reflected in the increased attention given to the body in tourism (Jokinen and Veijola 1994, Crouch 1999, Edensor 2001, Franklin 2001, Saldanha 2002, Obrador-Pons 2003, Bærenholdt et al 2004, Andrews 2005, Pritchard 2007, Evers 2009) and the complex materialialities of tourism encounters (Gibson 2009). However, tourist studies remains a product of an academic climate that celebrates visual culture, as is testified by the popularity of Urry's (1990) notion of the tourist gaze and MacCannell's (1999) view on tourist representations. There is little space for the multiple sensualities of the body in the most popular academic accounts of tourism as they tend to emphasise visual forms of consumption at the expense of the sensual impoverishment and de-materialisation of tourism. As Jokinen and Veijola (1994, 149) pointed out back in 1994 '[t]he tourist has lacked a body because the analyses have tended to concentrate on the gaze and/or structures and dynamics of waged labor societies'. There is nothing wrong with accounts that emphasise the significance of the visual as tourism is a prime example of the 'ocularcentrism' that has historically dominated western modern culture (Jay 1994) and is still heavily reliant on the visual sense and more specifically the gaze, as the case of tourist photography demonstrates (Crang 1999). The problem with tourism theory is that with such an emphasis on vision comes what Paterson (2007, 6) calls 'the forgetting of touch', the neglect of the haptic and all the array of sensual dispositions and structures of feeling which are mobilised through it. There is still relatively little interest in tourist studies for the way we feel and perceive the environment as well as the formation and regulation of the 'sensorium' (Paterson, 2009, 71). It also comes a disdain for the conviviality of tourism as well the as embodied experiences of fun, pleasure and joy – which, as Yi-Fu Tuan (2005) explains, have a particularly close association with the sense of touch. The fields of leisure and tourism are prime examples of the predisposition of cultural studies 'to dismiss or debunk pleasure' (Classen 2005, 69) as tourist studies 'engages in the social reproduction of seriousness' (Franklin and Crang 2001, 14).

In an academic climate that continues to celebrate visual forms of consumption the beach remains largely invisible. There are relatively few examples of work that examine the cultural significance of the beach, the main exceptions being the work of Urbain (2003) and Lenček and Bosker (1999). Most of the works on cultures and practices of tourists still focus on specialist forms of usually high status tourism. The invisibility of the beach in tourist studies does not simply reflect a gap in knowledge, it is also a consequence of the way tourism have been conceptualised in the Social Sciences as a serious, solitary and visual activity. Dominant accounts of tourism struggle to grasp a space that is paradigmatic of what Howes (2004) calls the 'sensual logic of late capitalism', that is, the increasing prominence of sensory stimulation in consumer capitalism. The beach cannot be easily reduced to a visual logic, even when the exhibition of flesh is very important. As the three S's of tourism 'sun, sea and sand' suggest, the beach appeals to the sensual body more than the penetrating eye. Its popularity depends on embodied pleasures such as sunbathing (Obrador-Pons 2009), swimming (Corbin 1994) or surfing (Booth 2001, Evers 2009). However, the sensuality of the beach refers to displeasure as much as pleasure. Its detractors often refer to the pain of sunburn, the boredom of inactivity or the feeling of shame that goes with undressing the body. The beach is populated with a series of objects, technologies and artefacts such as sun creams, parasols, swimming costumes that have been purposely designed to modify the human sensorium, amplifying the sensual potentialities of this environment. Conviviality is also central to the culture of the beach as the name of the most famous beach in Britain – the Pleasure Beach in Blackpool – suggest (Bennett 1983). In the west the beach is convivial, ludic and ritual; 'The beach is spectacular', Urbain (2003, 7) explains, 'it is a theatre in which society unveils itself, lays itself bare (literally and metaphorically)'. It is a liminal and carnivalesque space (Shields 1991, Urbain 2003) where ordinary social rules are blurred and suspended and new embodied sensibilities are tested out.

Both sensual and convivial dimensions are generally missing from popular accounts of the beach. Films, postcards and novels such as *Captain Corelli's Mandolin* (Bernières 1995), *Robinson Crusoe* (Defoe 2009) or Garland's *The Beach* (1997) tend to portray the beach as an Eden, as the ultimate garden of the world, an earthly paradise where everyone wants to return, 'a place so benign and beautiful and good' – Cronon (1996, 37) explains, 'that the imperative to preserve or restore it could be questioned only by those who ally themselves with evil'. The same is true for commercial accounts of the beach as the tourism industry makes a living by selling paradise for the price of the holiday. The presence of people's bodies clashes with such a powerful Edenic narrative, which invariably portray the beach untouched and picturesque as a place removed from civiliation, a place without bodies and tribes which is always in danger of mass invasion and 'collective rape' (Urbain 2003, 13). In these accounts the potential inhabitant of the beach is seen as an 'incurable nostalgic' (Urbain 2003, 10) that plays at being Robinson Crusoe, colonising the garden of Eden 'in the company of not too many Fridays, safe from the Savages' (Urbain 2003, 14), hoping that not many people will disrupt his corner of Eden. Such disembodied narratives stand out against a backdrop that actually celebrates the body as the beach is a place of sensuality and body work, 'very much the site of the making of the modern body' according to Löfgren (1999, 224). In Western societies the beach is also intimately linked to the changing cultural attitudes towards nudity (Barcan 2004), the complex association between sunlight, bodies and health (Carter 2007) and, of course, a body aesthetics that emphasise bodywork, tanning and hairless skin.

Furthermore, the beach is one of the few environments that break with the contemporary tendency to restrict tactile stimulation. On the beach we actually touch the ground and lie on it with our body without the mediation of artificial skins (Howes 2005, 28). There are some occasions where these elements surface and the beach is pictured full with bodies. However, when this happens beach bodies tend to adopt *animalistic* traits as tourist are redefined as *'turistas vulagris'* (Franklin and Crang 2001, 8) with voluptuous bodies gathering in herds. A good example of this is the 'saucy' comic postcard tradition of the English seaside resorts, which focuses on beach bodies and beach situations. These bodies are generally associated with mass tourism and function as a negative marker of class, thus sanctioning a well rehearsed set of ideological and social distinctions (Obrador et al 2009, 5). Rather than re-sensitising the beach, these alternative accounts ultimately confirm the Edenic tale of paradise lost through human interference, thus reproducing the western opposition between reason and sensuality, mind and body, to which 'the brute physicality of touch' (Classen 2005, 5) is associated.

This chapter is concerned with the sensuous dispositions that are mobilised on the beach. I am interested in the *making* and *sensing* of the beach, in how it is actually sensually and culturally experienced. The focus of this chapter is on the socio-historical construction of the 'sensorium' (Paterson 2009, 71), which has transformed a marginal space 'into a thriving, civilized, pleasure – and recreation-orientated outpost of Western life where so many sybaritic impulses of culture have been indelibly concentrated' (Lenček and Bosker 1999, xxi). The way we collectively and individually perceive and understand our environment is contextual and culturally learned as the 'sensory envelop of the self' (Paterson 2009, 71) adjusts in response to changing cultural attitudes, for example on nudity; the interaction with the physical environment, the 'sun, sea and sand'; and the mediation of technologies such as sun creams. This chapter responds to these concerns by focussing on the significance of the haptic sense on the beach, which is understood in a wider sense as 'an orientation to sensuality as such' (Paterson 2007, 2) as the haptic system of perception exceeds the scope of cutaneous sensations. Thinking of the beach in terms of touch provides a way to write meaningfully about the embodied experience of the beach as it makes space for sensuality, enjoyment and pleasure. In emphasising touch, I want to avoid producing a 'mortuary geography drained of the actual life' (Crang 1999, 248) that neglects texture, sensation and movement, that is, a geography that disregards what is important, the lived as lived, or as Wittgenstein put it 'the practical rough ground of life and thought' (quoted in Harrison 2002, 489). Touch provides a valuable route for bringing back life to the beach as it offers the opportunity to develop insights into its sensual richness, its embodied pleasures and displeasures as well as the interlacing of the body and the environment. It also provides a valuable route to re-materialise the beach, as metaphors of touch emphasise grasp and involvement rather than detached observation. The sensual and ludic dimensions of the beach are not impediments to knowledge but some of the main clues we have to make sense of this special environment. The beach matters because of its ludic and sensual qualities and not in spite of them.

This chapter draws on a style of thinking that emphasises the significance of haptic knowledges and geographies, the main examples being the work of Paterson (2004, 2006, 2007 and 2009), Vasseleu (1998), Classen (2005), Hetherington (2003) Bingley (2003) and Levinas (1997). Three basic ideas inform the way I bring touch into play. The first idea is that modalities of touch provide a framework of proximity, openness and intersubjectivity that defies oculocentric epistemologies wherein a detached subject contemplates the world from an outside point of view (Vasseleu 1998, Hetherington 2003, Paterson 2004 and 2006). Touch produces a distinctive form of knowledge that destabilises the neat, gendered order of the sensible associated with the tourist gaze. Whereas visual epistemologies tend to emphasise a dispassionate and idealistic form of knowledge, touch enables a more performative and embodied relation with the environment. The haptic system of perception is a site of relationship and affect, not of separation and atomism. Secondly the haptic system of perception is more than a simple receptor of skin sensations of pressure and contact, it is 'an orientation to sensuality as such' (Paterson 2007, 2) as well as 'a sense of communication' (Paterson 2007, 1), which is internally felt as much as it is externally orientated. It incorporates a whole range of sensations distributed throughout the body including kinaesthesia (the sense of movement), proprioception (a sense of bodily position) and the vestibular system (a sense of balance) which add to more *standard* cutaneous sensations. This chapter builds on an expanded notion of the haptic that emphasises the expressive and sensual dimensions of tactility. This is evident for example in the use of Levinas' (1997) notion of the caress to illustrate the more elusive and receptive forms of touch that are mobilised on the beach. An expanded haptic framework of analysis is more attuned to embodied pleasures and dispositions that animate the beach. Last, but not least, touch is more sensitive to the material textures of the beach. Touch reveals our 'withness with things' (Paterson 2007, 93), our being in an excessive material world. It also reveals the singularity of things, their uniqueness and irreplaceableness (Vasseleu 1998). Central to touch is an idea of confirmation, authenticity and presence, which Hetherington (2003, 1941) following Josipovici calls *praesentia*. In touch, things are significant in and of themselves, thus breaking with the dematerialised visualities of the optical experience.

While emphasising touch this chapter does not intend to replace the objectivity of vision with the proximity of touch, but to open up the Edenic account of the beach to something more performative. Neither is it my intention to find a shortcut to a more *pure* body that is supposedly free from cultural dressings. My aim is to reposition the beach within a tactile and performative space. In so doing, I follow the example of Wylie (2002), who takes inspiration from Merleau-Ponty's (1968) writing on phenomenology. In this view, vision is returned to a tactile sensible world, reinvesting light with carnal significance. By emphasising the distinctiveness of touch, however, this paper departs from Merleau-Ponty's phenomenology, which conceptualises vision and touch as reversible sensualities (Paterson 2004, 169, Vasseleu 1998, 67). Reasserting the tactile over the economy of the visual allows me to re-invest the beach with texture, sensation and enjoyment. While this chapter focuses on the more than representational side of the body, it does not locate it outside power and discourse, in a romantic way. The sensual dispositions that I examine are not conceived as originally prediscursive elements, but they are an effect of complex technologies of power and discourse, where the representational and the sensual are mutually constitutive. The focus of the chapter is on the specific socio-historical constructions of sensuality through which western tourists dwell on the beach.

My research has identified three patterns of haptic experience that have sedimented on the beach. A highly manipulative and playful version of touch enacted in the building of sandcastles coexists on the beach with the more elusive tactile experience of sunbathing, which centres on the cultivation of embodied qualities of receptivity and stillness, and the more kinaesthetic practices of swimming and bathing in the sea, which builds upon a sense of movement, pressure and temperature. The chapter is divided in three sections, each of which corresponds with one of the patterns of haptic experience that I have identified. With a focus on sandcastles the first section is concerned with the haptic experience of the sand. The section develops insights into the more ludic and performative dimension of the beach as well as the tangible materialities of the haptic experience. Building sandcastles is a highly performative activity that uses the manipulative qualities of sand to disclose virtual worlds. The second section focuses on sunbathing and examines the haptic relation with the sun. Sunbathing has been often associated with the visual sense, however it is first and foremost a haptic activity that builds upon the reception qualities of touch. In order to illustrate these more elusive forms of (non)-touch, I draw upon Levinas' (1997) notion of the caress. The final section is concerned with the practice of swimming and bathing in the sea. It develops insights into the more kinaesthetic modalities of touch, an aspect lacking in my previous work on the beach. The sensuous dispositions that these activities enact are the most somatic and internally orientated on the beach. The practices of swimming and bathing in the sea are predicated on the senses of movement, bodily position, balance and temperature rather than straightforward cutaneous touch.

This chapter is part of a wider project on the beach that foregrounds the significance of haptic knowledges and which includes work on nudity (Obrador-Pons 2007) sunbathing and the building of sandcastles (Obrador-Pons 2009). This project has its origins in ethnographic research that I conducted in the summers of 2002 and 2003 in three different beaches in the island of Menorca. My research was based on participant observation and semi-structured interviews. On each beach I spent five hours a day for at least a week, always on my own. As well as keeping a diary, I conducted a total of 55 interviews, involving approximately 140 people, all of who were tourists, and focused primarily on the more sensual and emotional aspects of the experience of the beach. This chapter builds on that project by developing an auto-ethnographic (Crang 2003, Picart 2002) stance which was not present in my previous writings. This stance is evident in particular in the section on touching the sea, which relies primarily on my own subjective experiences of the beach. By focusing on my own experiences as well as that of others, I respond to one of the biggest challenges of my research, how to do participant *observation* without appearing *voyeuristic* or *narcissistic,* that is, how to look into the sensuous culture of the beach without adopting a dispassionate and detached point of view. This chapter is neither just about other people's bodies nor a self-absorbed reflection on my own practice and positionality as researcher. What I am doing here is using myself 'as the instrument of research' (Crang 2003, 499), learning through my body's responses and situations, opening it up to a series of sensuous dispositions which compose the tourist experience of the beach. As a regular visitor of the beach, this is not as much a question of developing an insider perspective of an alien sensorium as a conscious effort to rationalise the sensory culture in which I dwell.

Touching the Sand

The beach takes shape and gains expression through a series of embodied, everyday social practices, most of which call attention to the significance of haptic knowledges and sensualities. The building of sandcastles is one of the most characteristic set of social practices through which the beach takes shape and gains expression. 'Through the sandcastle', Bærenholdt et al (2004, 3) explain, 'the space and materiality of the beach is domesticated, occupied, inhabited, embodied. The sandcastle transforms the endless mass of white, golden, fine grained or gravelled sand into a habitat; a kingdom imbued with dreams, hopes and pride'. Such a process of domestication – of 'transformation of the beach into a social space' (Bærenholdt et al 2004, 3) – is evident in Figure 2.1, where we see a large, rectangular castle with a moat, two rows of walls and a number of round towers. The family who built it can be seen working away in the background absorbed in the task of expanding their castle, under dad's careful direction. While a good example of tourist family photography (Haldrup and Larsen 2003), this picture also speaks of the performative and haptic character of the beach. The family in the picture inhabits the beach by playing with the sand, modelling ephemeral constructions with their hands and the bucket and spade. In foregrounding our practical involvement with the beach, this picture expose the beach as a skilful and technical accomplishment thus confirming that 'humanity is technical from the start' (Thrift and Dewsbury 2000, 419). Instead of viewing the beach as pre-figured in advance, as the product of external cultural process of construction, it situates its emergence as a tourist space in the process of appropriation, that is, in the actual 'doings' and 'dreamings' of tourists on the beach. 'It is the corporeal and social performances of tourists that make places touristic' (Bærenholdt et al 2004, 2). Such process in turns depends on vast and

Figure 2.1   A family working on a castle

Source: author photograph

extensive networks of infrastructure, narratives and public policies that make it possible.

The pattern of haptic experience that is enacted in the building of sandcastles privileges the manipulative tactility of the hand over other haptic articulations of sensuality. The building of sandcastles is predicated on highly performative skills such as digging, holding, moving, carrying, wetting the sand and making shapes. The beach is sensually and culturally experienced in an affective and embodied fashion through the touching hand that transforms the fine grains of sand into splendorous constructions, drawing together sand, dreams and flesh. This pattern of haptic experience corresponds with Western conceptions of touch. In Western cultures vision has been intellectualised and therefore disembodied. It is understood as an extension of the mind, 'distant and even deceitful' (Paterson 2007, 2). Touch, in contrast, is the 'primal sense' (Bingley 2003, 334) and is invariably connected with the immediacy and corporeality of everyday life. As such the experiences that it affords 'seems more intimate, reassuring and proximal' (Paterson 2007, 2). Further, as Hetherington (2003, 1934) explains '[t]ouch in our culture assumes a form of knowledge that is often more proximal than distal in kind (…). Proximal knowledge is performative rather than representational. Its non-representational quality is also context-specific, fragmentary and often mundane'. Instead of recreating an epistemological distance between subject and object, the tactility of sandcastles stresses a feeling of doing and practical engagement as well as a sense of proximity

Figure 2.2   Doing 'major engineering works'
Source: author photograph

and tangibility. Like in the case of hunting and angling (Franklin 2001), it promises a highly sensualised, intimate and performative relation with beach natures.

Building sandcastles is performative not only in the sense that it is tactile, but also in the sense that it is creative and processual. While enacting a highly performative set of haptic skills, building sandcastles also functions as a mode of disclosure and intervention, of actualisation of the virtual. A difference needs to be established here between what Thrift and Dewsbury (2000, 416) call, drawing on Deleuze, the 'realisation of the possible' and the 'actualisation of the virtual', that is, between an understanding of performance as the reiteration of pre-existing social forms and an one that is concerned with unfolding new possibilities and the creation of affects. The point of building a sandcastle is not in the mimetic reproduction of a pre-formed design as that would mean that nothing new is created. The point is in the actual modelling of sand that facilitates the access to new layers of meaning and experience and enables the invention of other spaces. The building of sandcastles 'does not operate through resemblance or representation, but by generating difference, divergence and creation' (Thrift and Dewsbury 2000, 416).

Building sandcastles can also be aligned with dance as an example of play in the sense that Radley (1995) and Thrift (1997) use it. Play is neither an act of inscription nor an act of liberation but a *performative experiment* (Thrift 1997, 145). Play builds upon the expressive capacity of the body – in this case the elusory qualities of the touching hand – to conjure up virtual 'as-if' worlds, in other words, 'ways of being that can become claims to "something more"'

(Radley 1995, 13). This notion of play and playfulness is different from the one Urry (1990) associates with the ironic and cool figure of the post-tourist. Play here does not suggest inauthenticity, superficiality or immediate gratification, but it is, paraphrasing Bingley (2003, 331), a *potential space* that has a mediating and relational function. Through play a 'facilitating environment' is created that both stimulates expression and enables the exploration of the self in (relation to) space. Play is, according to Bingley (2003, 331), 'a "day-dream-like" space where fantasy, dreams and the real world can meet, and sense be made of the nature and interrelationship of inner and outer objects and world'. Through the manipulation of the sand alternative ways of being are configured, new geographies are opened out, new connections and assemblages are tested and some meanings and utopias are spatialised. The beach with its sandcastles is a virtual and performative space unfolded through play.

The pictures that accompany this section are frozen instants of what is a complex set of performances. They conceal the dimensions of time, process, transformation and deterioration, the fluid and changing character of sandcastles. In this way, they reproduce the conceptual displacement that according to Massumi (2002, 1) characterises contemporary cultural theory, the displacement of '(movement/sensation)-change'. However, some constructions have a stronger performative character than others. This is the case in Figure 2.2 where we can see a complex system of pits, walls and canals, 'major engineering works' according to one of my respondents. In fact, it can hardly be called a construction, but simply a collection of piles and holes in the sand. It is an uncompleted product, an unfinished performance, a story made of sand. The purpose of making holes is to dig deep into the water underneath; the walls are made to protect the hole from the erosion of the sea. First the children will build a hole and then a wall to separate their new *swimming pool* from the sea. Once this task is completed they will open a canal in order to let the seawater fill the hole. However, a few minutes later a wave will break the wall destroying their constructions. Excited with the challenge the sea poses, the children will often rebuild bigger walls to protect their hole from the sea. Such a performative sense tends to be less intense in most of the adult constructions. The sandcastle pictured in Figure 2.3, for example, made by an adult, does not have the temporal dimension that children's castles have. Although the actual process of building a castle is still what matters most, the emphases of adult sandcastles tend to be on inscription rather than performance, on the sculptural form rather than the story.

Thinking in terms of touch makes it possible to re-emphasise the material textures of the beach, along with a sense of enjoyment and enchantment. It has been widely assumed that re-emphasising the material leads to a more solid sense of reality. In the case of the beach, however, an openness to materiality exceeds any invocation for a more grounded and empirical analysis. This stance echoes recent debates in human geography (Anderson and Tolia-Kelly 2004, Anderson and Wylie 2009) that call for more complicated and experimental figurations of the material that challenge the traditional division between culture and matter, word

Figure 2.3   A crocodile
Source: author photograph

and world. An understanding of the material as 'inert blank or radical outside' (Anderson and Tolia-Kelly 2004, 672) is, therefore, rejected in favour of a focus that emphasises the liveliness and singularities of the multiple materialities at play on the beach. The pattern of experience that touch produces stresses the singularity of things, their uniqueness and irreplaceableness. On the beach the objects, materials and landscapes that we encounter tactilely are significant in and of themselves. Touch reveals our 'withness with things' (Paterson 2007, 93), our being in a vibrant, quirky, overflowing material world. A sense of singularity is key to Vasseleu's understanding of touch. 'The non-substitutability of bodies', Vasseleu (1998, 71) argues, 'is experienced in the divergence of touch, but this divergence cannot be recovered in the reversal between seer and visible. Nothing equivalent fulfils/takes the place of the invisibility of touch'. Central to touch is a sense of confirmation and authenticity of the material; as Hetherington (2003, 1941) explains 'we touch something to confirm that there is there, that it feels like this, that our eyes do not deceive us'. Touch, as Bingley (2003) points out, is also capable of evoking memories, meanings and utopias, which are invariably linked to childhood. On the beach something absent – a dream, a fantasy or a feeling – can

Figure 2.4   A castle

Source: author photograph

attain a presence simply through the manipulation of sand. Drawing on the work of Josipovici, Hetherington (2003, 1937) identifies these heterogeneous qualities of touch, which are fundamental to the building of sandcastles, as *praesentia*.

Building sandcastles is, therefore, about making things present and tangible. The beach is a place in which it is literally possible to make virtual worlds actual and touchable. On the beach inert masses of sand are transformed into solid material configurations that evoke a wide range of aspects generally related with childhood play experiences, ideas and fantasies. Underpinning sandcastles there are a myriad of secret stories of knights, princesses, soldiers, invaders, and battles. On some occasions these stories build upon the latest films or the most popular children's book, perhaps *Harry Potter* or *Star Trek*. On other occasions, as it is probably the case in Figure 2.4, they enact a legendary account of the medieval castle, one which has been played out many times before. There are sandcastles which are a direct attempt to build a mimetic reproduction of a building, an object or a creature. This is the case of Figures 2.3 and 2.5 where we can see a crocodile and a boat. These examples call attention to the fact that the beach is not an original pre-discursive site outside the realm of the social, but is itself an effect of complex technologies of power and discourse, where the representational and the sensual are mutually constitutive. The textual and representational elements shaping a sandcastle are integral to the beach as they confer meaning and inhabitability to the sand.

The plasticity of the sand is what ultimately makes possible for something absent to attain a presence on the beach. The sand is a very manipulative and

Figure 2.5   A boat
Source: author photograph

malleable material with endless potentialities for playing. Children can bury themselves, try a handstand or build all sorts of constructions. Few other materials have the plasticity and fluidity of the sand. Such a connection between play and the material qualities of the sand is highlighted in the work of Bingley (2003. See also Bingley and Milligan 2007) on sensory experience and landscape as she incorporates sand play activities in her research. Bingley found that modelling work facilitated the exploration and expression of the relationship between self and landscape. For Bingley (2003, 336) the sand is 'a "blank canvas" upon which participants could freely express, either their immediate sensory experience, and/ or the emergent past memories and meanings in relation to Self and Other". The plasticity of the sand enables both adults and children 'to reconnect with play' and 'to get in touch with touch' (Bingley 2003, 334–5), that is, to engage with generally hidden aspects of the self and the landscape. It brings to light how the beach has been spatially constituted to facilitate access to other dimensions of social experience which are not readily available through cognitive processes.

In the building of sandcastles, the particular textures and sensualities of the sand matter. If the grain is too thick it is difficult to give consistency to the castle, whereas if it is too thin the grains stick to the body. The material textures of the sand with its inherent properties and potentialities are not passive spectators but, in fact, a very important part of the experience of the beach. Texture constitutes a tourist attraction in itself as people dwell upon the mundane sensual qualities of sand. This is evident in Menorca where due to its heterogeneous geologic composition there are a great variety of sands. The south coast is dominated by limestone and the resulting sand is generally white, fine and quite sticky. It is good sand for building castles and confers a much loved tropical look, although it often contains seaweed. The geologic formations that are predominant in the north are older and more heterogeneous. There are even significant concentrations of black clay. As a result the sand is normally thicker, less sticky and of different colours. In some places it is rather yellow and clay-like, while in others it has tints of black and grey. The presence of shells and other marine animals is also significant here. Each beach then has a little treasure hidden in the sand that people like to collect. These treasures are often used in the building of sandcastle to recreate vegetation or to coronate towers. All these textures are central in the building of sandcastles and the experience of the beach more generally.

Notwithstanding the singularities and potentialities of the sand, the material configurations of sandcastles are intrinsically precarious and ephemeral. As Baerenholdt et al (2004) explain they are evanescent and fugacious creations that the waves, the wind and people will vanish away. Tunnels and towers may collapse as the sun shines, and the wet sand dries up, and the texture changes. The rising tide may cause water to penetrate the ramparts surrounding the moat and undermine the fortifications of the castle. The work of erosion and sedimentation can alter the sandcastle slowly or with sudden ruptures. It changes their appearance and inspires new reconstructions, a submerged moat may inspire the construction of a channel to the sea, a collapsed tower affords space for an enlarging the stronghold (Baerenholdt et al 2004, 6). The textures of the sand are multiple, concrete and vibrant, but they are never fixed, always changing and elusive. Nor are sandcastles, whose physical presence is short and fugacious, always shifting between an actual and a virtual state. The sand is an autonomous and fluid materiality empowered with movement. All the while we play with the sand, the sand too is playing with us.

Touching the Sun

Sunbathing is probably the most popular tourist activity in the Mediterranean. Most tourists come on holiday to places like Menorca at the hottest time of the year to spend hours on end lying on the beach covered in sun cream to enjoy the bliss of the sun. The heliocentrism of contemporary tourism is a relatively modern phenomenon that emerged according to Carter (2007) in the first years of the twentieth century when a positive association between sunlight and bodies was established. This new association brought to an end the routine isolation of bodies from sunlight as well as the aristocratic codes of beauty that invariably privileged pale white skin. Sunbathing has recurrently been associated with the visual sense – with good-looking tanned bodies. There is certainly an element of that; however a visual framework does not exhaust the experience of the sun. As important as the tan, the visual spectacle of coloured bodies, is getting the tan, that is, the touch of the sun on the skin. Sunbathing therefore also calls attention to the significance of haptic knowledges and sensualities on the beach, however, it suggests a different order of the sensible to that of sandcastles as it builds upon more receptive modalities of touch in particular a cutaneous awareness of temperature. The pattern of haptic experience associated with sunbathing is closer to the notion of the caress than the idea of grasp. By turning my attention to sunbathing I want to show how the beach not only functions as a playful space but also as pseudo-contemplative space. The highly manipulative version of touch that is enacted in the building of sandcastles coexists with more elusive tactilities that are centred on the cultivation of the embodied qualities of receptivity and stillness.

Levinas' notion of the caress provides a framework for thinking the pattern of haptic experience associated with sunbathing. There is no prehension or manipulation involved in sunbathing, but a form of touch that like the caress eludes the grasp. According to Levinas (1997, 89):

> The caress is a mode of the subject's being, where the subject who is in contact with another goes beyond this contact. Contact as sensation is part of the world of light. But what is caressed is not touched, properly speaking. It is not the softness or warmth of the hand given in contact that the other seeks. The seeking of the cares constitutes its essence by the fact that the caress does not know what it seeks. This 'not knowing', this fundamental disorder is the essential.

Central to Levinas' notion of the caress is a sense of the elusive. Caressing implies a loss of perspective, 'a withdrawal form the harshness of the light' and 'a losing sight of touch as sensation' (Vasseleu 1998, 105–6). The caress does not work through possession, incorporation or assimilation but through relation, where alterity remains intact as other; it is 'an obsession with a non-negatable difference which persist in the absence of light' (Vasseleu 1998, 105). Being intimately linked with the caress, sunbathing also involves an impossibility of translating the experience into words, knowledge and power. The sun seeker does not intend to master the environment, but simply to find delight in the relation with it. The sense of elusiveness associated with sunbathing contrasts with the sense of tangibility and textured manipulation that are central to the building of sandcastles. Sunbathing entails a much more quiet, peaceful and sensual relation with the natural elements that resist incorporation. Allowing the sun onto the body, warming the skin and feeling relaxed and sleepy are some of the main delights people find in sunbathing. Sunbathing is, first and foremost, an act of openness to the sun as well as an awareness of temperature.

Conventional verbal methods do not provide full access to the sensually rich experience of sunbathing. It is a difficult activity for people to rationalise into words and it can only be explored auto-ethnographically through embodied experimentation. However, the interviews I conducted on the beach did afford some clues that hinted at a pattern of haptic experience that exceeds manipulation and which has a lot in common with Levinas' concept of the caress. Although nobody directly related sunbathing with touch, the majority of people that I talked to described sunbathing in haptic sensual terms as a pleasurable feeling. This is the case for Ruth from California[1] 'I love the feeling of the sun on my skin'– she emphasised – 'it's just the best!'. Very often this feeling is described as relaxing. Take, for example, the case of Javier from mainland Spain, who elaborates on the relaxing qualities of the sun. 'A bath of sun is very relaxing. I love it'. He even compares himself with a sun-loving creature 'I am like a lizard. I can stay here until the sunset'. In so doing, Javier links sunbathing with the cultivation of embodied qualities of receptivity and stillness. Most of the observations I collected did not come as clear and rational statements, but simply as general appellations to sensuality. Sometimes these appellations are punctuated by emotion and even a metaphysical discourse. In the following quotation Janet from Paris describes sunbathing primarily as an awareness of temperature and she qualifies such feelings as relaxing and almost spiritual: 'I like being in the sun because when I just close my eyes and I feel I'm getting very warm I think … I feel I'm getting more and more relaxed and I think it's good for my body and for my soul and for everything'. In this quotation sunbathing is depicted as a receptive and semi-contemplative state which requires the closing of the eyes, a withdrawal from light. Here sunbathing is neither an empty nor a passive state, but one which is excessive of meaning and life and has to be actively cultivated.

By taking touch beyond manipulation we can argue towards a stronger notion of the tactile and tactility in the sense that Irigaray gives to these concepts, which is according to Paterson (2004, 171), 'communicative, extra-linguistic and intersubjective openness to the other'. The caress establishes a distinctive relation with the environment which is more intimate and fluid. When sunbathing we do not feel our bodies as having distinct boundaries with the environment, rather we feel a constant fluid exchange between bodies and nature. The caress is a site of relationship and affect in which there is no clear distinction between the subject and object as clear and coherent entities; 'our bodies extend into things and they extend into us' (Paterson 2004, 175). This modality of touch also contains clear erotic connotations, some of which are evident in the practice of sunbathing as for example in rubbing oil on the skin. However, the caress should not be associated with passivity but with the embodied skills of receptivity and stillness. Sunbathing is about learning to be affective, open to the experience of touching and entails an actively passive communication; '[o]ne has to be open to the world in order to take it in' (Paterson 2004, 170). Sunbathing is about stillness – unlike the building of sandcastles it lacks a strong sense of the creative and the performative – however it is not an empty moment. Caressing does not refer to a lack of palpability, but to a moment overfull of action that is excessive. Sunbathing brings into contact with elusive presences lying beyond but passing through the body.

---

1   The profile of Ruth (not her real name) is highly unusual in Menorca, which attracts very few North American or Asian tourists. Ruth was visiting a friend from Barcelona that had a holiday home in Menorca.

Sunbathing suggests a different pattern of haptic experience to that of sandcastles. Instead of prehension or manipulation, sunbathing builds upon much more receptive and elusive modalities of touch, which have so much in common with Levinas' (1997) notion of the caress. However, the extent to which this connections holds true is a matter for further discussion particularly in relation to the seeming paradox that sunbathing involves a pursuit of light whereas the caress involves a withdrawal from light. Perhaps the case of sunbathing is pointing to a different more sensual relation with light that challenge Western associations with visibility and rationality. Another element of discussion is the extent to which sunbathing does share with Levinas' notion of the caress an erotic dimension.

Touching the Sea

While tourism in Menorca is undoubtedly heliocentric, it is the sea and not the sun that tourists tend to value the most. 'I come to Menorca in search of the sea', asserted one of my respondents. Menorca is especially renowned for its pleasant, warm and transparent waters, which also constitute the central theme of most tourist brochures. The appeal of the sea stems in large measure from its haptic qualities. Of the three 'Ss' of beach tourism the sea is the element with the strongest haptic orientation, not least in Menorca. The significance of the haptic qualities of the sea is evident in the way most of my respondents made use of a sensual language to talk about its appeal. In my fieldwork I repeatedly come across with assertions such as 'the waters are very appetizing'. These assertions suggest the importance of the sensual feel of the sea, for many as important as its look, if not more. The most appreciated qualities are temperature, transparency and stillness. 'The sea is so warm' – many of my respondents emphasised while pondering the qualities of the beach. 'Why come to a beautiful place like this and not enjoy the scenery and come close to the water?' – another of my respondents asked rhetorically. The relaxing and rhythmic sound of the waves was also highlighted by a number of tourists. These haptic qualities are relational in character as most of my respondents brought them into light when comparing Menorca with other destinations. Take for example the case of Ruth from California. She praised the pleasant waters of Menorca by contrasting them to the cold and rough waters of California. 'Well the waves are huge in California and it's a lot colder' – she explained – '[in California] you can swim maybe on a really hot day get in and get out but you don't want it, you just paddle around'. For Ruth this is different in Menorca where you can swim all summer long as waters are clean, tranquil and warm. In most of the comments that I collected concerning the sea there was, paraphrasing Paterson (2007, 6), a 'conceptual slippage between touching and feeling'. Haptic sensations concerning the sea were recurrently associated with particular feelings as in the case of Javier for whom the sea produced a 'sensation of freedom'; or Ramon who contended that 'the sea has depth, abstraction and calmness'; 'it is a projection with the horizon' he concluded. This is the case because the haptic is much more than a simple receptor of information, touch is a expressive and emotionally productive; it is, according to Paterson (2007, 2), 'an orientation to sensuality as such'.

While sunbathing is centred on the elusive tactilities of the caress and the construction of sandcastles builds upon highly manipulative and playful modalities of touch, aquatic activities are predicated mainly on the haptic senses of movement, bodily position, balance and temperature. The pattern of haptic experience enacted in the sea is first and foremost kinaesthetic. In the sea most leisure activities involve some form of movement, including swimming, surfing, diving, and sailing. Such movement takes place in a liquid environment with unusual conditions of pressure, density and temperature where the human body is not *naturally* comfortable. While the practice of swimming consists in moving the body through the water without drowning; the point of diving is the acceleration of the body in free fall. Scuba diving implies the movement of the body under the water whereas in the case of surfing is over the water in conjunction with the movement of the sea. What motivate these kinaesthetic practices is the thrill and ecstasy that moving the body in different and often extreme conditions produce. Drawing on the work of Stranger (1999), Franklin describes accurately the kinaesthetic thrill associated with surfing. According to Franklin (2003, 235) '[t]he intensity of the kinaesthetic exercise involved in surfing techniques in combination with the intensity of the challenge locks the surfer into what [Stranger 1999] calls "an ecstatic moment"'. The senses of proprioception (sense of bodily position) and the vestibular system (the sense of balance) are also very important. This is especially true in the case of surfing, which among other things plays with the destabilisation of balance and the experimentation with unusual positions of the body. In his work on masculinity, bodies and surfing, Evers (2009) thinks of surfing not so much as a form of colonisation of nature but as a close sensual relationship with an ever-changing environment which is in a constant state of flux. In riding waves, surfers establish a complex, dynamic assemblage with a fluid environment that 'never allows proprioceptic, kinaesthetic and sensual awareness to settle' (Evers 2009, 898). 'Surfers ride with waves, not simply on them. They are part of our bodies and our bodies are part of them' (Evers 2009, 898). The point of surfing is to be found, therefore, in the attunement of the body, including its position and balance, with the rhythms and textures of a liquid environment, resulting in 'a body that surfs' (Evers 2009, 898).

Haptic sensations of temperature also matter to the modern experience of the beach. Whereas back in the seventeenth century it was the feeling of cold water that mattered the most as therapeutic sea baths become increasingly popular (Lenček and Bosker 1999), today the feeling of temperature that is more attractive is to be found in warm waters. Cold waters are now a handicap as for example in surfing and diving, which often require the use of a wet suit to keep the body warm, while warmer waters tend to be felt as a bliss, not least in Menorca. During the hot summer months the experience of the beach in Menorca is organised as a succession of periods of warming up in the sun and cooling off in the water. It is often the freshness of the water that people remember of the most. As one of my respondents explained 'if you are too hot, you just have a swim, it's wonderful, getting out the cold water putting a towel on, it's wonderful'. In hot summer the freshness of the water can be a blessing.

What is interesting in the case of the sea is the interoceptive character of aquatic tactilities. The haptic sensualities enacted in the sea are the most somatic and internally orientated sensations enacted on the beach. Sea leisure activities do not pursue cutaneous contact as much as the feeling of the body in a different environment. The senses of movement, balance, temperature and bodily position on which these activities are predicated are internally felt sensations that report about the situation of the body. As Paterson (2007, 27) explains '[a] more contemporary psychology would treat somatic senses of proprioception, kinaesthesia and the vestibular sense as working synergistically, as the inwardly-orientated sensations necessary for feelings of embodiment'. With the sea we are confronted with a more comprehensive understanding of the haptic sense as not reducible to tactile sensation alone. 'Touch being a manifold of sensations', Paterson (2007, 27) explains, '"feeling" involves not only perception by touch... but also perception of our whole bodily state, involving interoception and somatic sensations'. The haptic experience of the sea is somatic and interoceptive primarily because of the specific materiality of the aquatic medium, which destabilises the human sensorium, as the body is not naturally comfortable in such unusual conditions. The sense of movement, bodily position and balance are so central to our sense of being in the world that most of us are unaware we have them. We only notice them when the *usual* human sensory is altered as in the case of the sea. Many of the leisure activities that take place in the sea play with such an alteration of the human sensory. This is often the case with children's aquatic games which tend to have an experimental character. In the water children are recurrently testing the body, experimenting with the sensual opportunities and challenges that the sea provides. Children experiment with their bodies often through simple actions such as touching the bottom of the sea, or putting the head under the water and opening the eyes, or floating by extending the body over the water, an action that in Menorca they call 'doing the dead'. An auto-ethnographical point of view is inevitable in a discussion on haptic knowledges which, as Paterson (2007) and Bingley (2003) point out, are usually indissociable from acts of memory. The vivid memory of my own childhood experimentations with the sea underpins much of what I have to say here. I still remember when. aged around 5 or 6 years old, how I learned to dive with my head down. I spent all the summer up and down on the rocks trying and testing my new skill. A sense experimentation with the sea was also evident in the people that I interviewed. When I asked about one of my respondents what they do on the beach she replied 'me ensuring that these two [referring to her husband and her son] are not drowning. And them, trying to drown themselves'. Father and son were enjoying the waves, playing with the movement of the sea and this was proving stressful for the wife.

Figure 2.6   Carrying an octopus in my hands

Source: author photograph

In the sea somatic and interoceptive aquatic tactilities coexist with more ordinary cutaneous forms of touch. Central to the experience of the beach is a desire to touch and feel certain textures. The sea is full of exciting and distinctive textures that invite us to look and touch including the soft white sands that confer a much desired tropical look to the beach, the spiky sea urchin hidden in the gaps of a rock and the squidgy creatures that populate more rocky areas. Figure 2.6 is a good illustration of such desire to touch. This is a picture of myself when I was five years old with an octopus in my hand that I managed to catch after sticking to my leg while playing in the sea. The picture suggests the excitement of holding a sea creature in my hand. Straughan (2010, 7) identifies a similar desire to touch in scuba diving: '[t]he direct refusal of touching obscures what is a powerful desire to touch and feel certain textures proffered by the aquatic world'. The things that divers wish to touch are often 'those very things that they would otherwise strongly argue against touching', such as coral (Straughan 2010, 7). However, not all textures are pleasing and inviting and certainly not everybody feels the same desire to touch. Some aquatic textures can be a source of embodied anxiety. A especially controversial texture in Menorca is the presence of dark coloured remains of aquatic vegetation along the shores. These belong to the *oceanic posidonea*, a Mediterranean endemic marine plant which forms authentic prairies across the sea bottom. The embodied instinct of many people is to associate this plant with pollution. However, its presence is the best guarantee of good health for the Menorcan beaches, as it is an important habitat for many species and contributes to the 'tropical look' of many of the beaches. Another controversial texture are the presence of jellyfishes on the shore. These creatures are neither particularly ugly nor move in an intimidating way as octopus do, but cutaneous contact with a jellyfish can burn the skin. Their presence in the seas has become a recurrent problems for tourism. Such desires and anxieties are a powerful reminder of the close association of touching and feeling. Haptic sensations do not appear in isolation but they are always entangled together forming a dynamic and contingent assemblage of sensations, objects and feelings.

Conclusions

While a somewhat geographically marginal space, the beach occupies a special place in the Western world. Countless mythological dramas, including various narratives of the Earth's origin, have been staged at the point where the land meets the sea, and the beach has been recurrently imagined as an Eden, the setting of paradise as well as 'nature's most potent anti-depressant' (Lenček and Bosker 1999, xiv). The beach is also the backbone of many tourism economies. The combination of sun, sea and sand has transformed the derelict economies of many peripheral regions, including the Mediterranean.

The beach then is as significant is cultural as much as it is economical (Obrador-Pons et al 2009), however, it has been largely invisible to tourist studies. Paradoxically this is partially due to the disciplinary attempts to seize its significance, which has led tourist studies to systematically disregard what is actually more important, the sensual and ludic elements that make the beach so special. Stripped of sensuality, enjoyment and conviviality the beach may be seen as a casualty of both the obsession of tourist studies with respectability and more generally the sensual poverty of the social theory. This chapter responds to these shortcomings by foregrounding the significance of touch on the beach. In so doing, the driving concern has been the development of framework of analysis that is more attune to the embodied pleasures and material textures of the beach. Drawing on ethnographical research in the island of Menorca, I have developed insights into some of the main articulations of haptic sensualities that are mobilised on the beach. I have identified three patterns of haptic experience, firstly on the beach a manipulative and playful version of touch enacted in the building of sandcastles. This coexists with the more elusive tactile experience of sunbathing, which centres on the cultivation of embodied qualities of receptivity and stillness. Thirdly, there is the more kinaesthetic practices of swimming, which builds upon a sense of movement, pressure and temperature.

In examining the sensuous dispositions that are mobilised on the beach, this chapter goes to the very heart of tourism economies as it develops insights into the sensual logic of beach-based recreation. The chapter responds to the belief that only by attending the articulations of practice and structures of feeling, is it possible to explain the mass appeal of the beach. The so often neglected senses of touch are not a minor element of the sensual cultures of the beach, but actually one of its most fundamental components. The playful character of the beach comes with the manipulative hand, its contemplative and sensual qualities with the more receptive modalities of touch and the thrill that is fundamental to so many aquatic activities with kinaesthesia. The haptic, the emotional and the cultural dimensions of the beach are indistinguishable and mutually constitutive. By exploring the rich sensualities of the beach, this chapter has also brought to the fore the significance of this liminal spaces to the sensual economies of Western societies. Haptic sensual qualities matter to the beach as much the beach matters to the socio-historical construction of the sensorium in the West. There is a need to take seriously the ludic and the sensual character of Western societies, not as a sign of weakness, but as major sites for stimulation and enchantment. The three 'Ss' of tourism are not simply an expression of superficiality and hedonism, but an evidence of how the contemporary world still inspires deep and powerful attachment.

References

Anderson, B. and Tolia-Kelly, D. 2004. Matter(s) in social and cultural geography'. *Geoforum*, 35(6), 669–74.

Anderson, B. and Wylie, J. 2009. On geography and materiality. *Environment and Planning A*, 41(1), 318–45.

Andrews, H. 2005. Feeling at home: Embodying Britishness in a Spanish charter tourist resort. *Tourist Studies*, 5(3), 247–66.

Bærenholdt, J.O., Haldrup, M., Larsen, J. and Urry, J. 2004. *Performing Tourist Places*. Aldershot, England: Ashgate.

Barcan, R. 2004. *Nudity: A Cultural Anatomy*. Oxford: Berg.

Bennett, T. 1983. A thousand and one troubles: Blackpool Pleasure Beach, in *Formations of Pleasure*, edited by F. Jameson. London: Routledge.

Bernières, L.D. 1995. *Captain Corelli's Mandolin: A Novel*. London: Vintage.

Bingley, A., 2003. In here and out there: sensations between self and landscape. *Social & Cultural Geography*, 4(3), 329–45.

Bingley, A. and Milligan, C. 2007. Sandplay, clay and sticks: Multi-sensory research methods to explore the long-term mental health effects of childhood play experience. *Children's Geographies*, 5(3), 283–96.

Booth, D. 2001. *Australian Beach Cultures: The History of Sun, Sand, and Surf.* London: Frank Cass.

Carter, S. 2007. *Rise and Shine: Sunlight, Technology and Health.* Oxford: Berg.

Classen, C. 2005. *The Book of Touch.* Oxford: Berg.

Corbin, A. 1994. *The Lure of the Sea: The Discovery of the Seaside in the Western World, 1750-1840.* Berkeley, CA: University of California Press.

Crang, M. 1999. Knowing, tourism and practices of vision, in *Leisure / Tourism Geographies: Practices and Geographical Knowledge*, edited by D. Crouch. London: Routledge.

Crang, M. 2003. Qualitative methods: touchy, feely, look-see? *Progress in Human Geography*, 27(4), 494–504.

Cronon, W. 1996. *Uncommon Ground: Rethinking the Human Place in Nature.* New York: Norton & Co.

Crouch, D. 1999. *Leisure / Tourism Geographies: Practices and Geographical Knowledge.* London: Routledge.

Defoe, D. 2009[1791]. *Robinson Crusoe.* Oxford: Oxford University Press.

Edensor, T. 2000. Walking in the British countryside: Reflexivity, embodiment and ways to escape. *Body & Society,* 6(3/4), 81–106.

Edensor, T. 2001. Performing tourism, staging tourism: (re)producing tourist space and practice. *Tourist Studies,* 1(1), 59–81.

Evers, C. 2009. "The Point": Surfing, geography and a sensual life of men and masculinity on the Gold Coast, Australia. *Social & Cultural Geography*, 10(8), 893–908.

Franklin, A. 2001. 'Neo-Darwinian leisures, the body and nature: hunting and angling in modernity. *Body & Society,* 7(4), 57–76.

Franklin, A. 2003. *Tourism: An Introduction.* London: Sage.

Franklin, A. and Crang, M. 2001. The trouble with tourism and travel theory. *Tourist Studies,* 1(1), 5–22.

Garland, A. 1997. *The Beach.* New York: Riverhead Books.

Gibson, C. 2009. Geographies of tourism: (un)ethical encounters. *Progress in Human Geography,* 33(5), 1–7.

Haldrup, M. and Larsen, J. 2003. The family gaze. *Tourist Studies,* 3(1), 23–46.

Harrison, P. 2002. The caesura: remarks on Wittgenstein's interruption of theory, or why practices elude explanation. *Geoforum,* 33(4), 487–503.

Hetherington, K. 2003. Spatial textures: making place through touch. *Environment and Planning A,* 35(11), 1933–44.

Howes, D. 2004. *Empire of the Senses: The Sensual Culture Reader.* Oxford: Berg.

Howes, D. 2005. Skinscapes: embodiment, culture, and environment, in *The Book of Touch*, edited by C. Classen. Oxford: Berg.

Jay, M. 1994. *Downcast Eyes: The Denigration of Vision in Twentieth-Century French Thought.* Berkeley, CA: University of California Press.

Jokinen, E. and Veijola, S. 1994. The body in tourism. *Theory, Culture & Society,* 11, 125–51.

Lenček, L. and Bosker, G. 1999 *The Beach: the History of Paradise on Earth.* London: Pimlico.

Levinas, E. 1997. *Time and the Other.* Pittsburgh, PA: Duquesne University Press.

Löfgren, O. 1999. *On Holiday: A History of Vacationing.* Berkeley, CA: University of California Press.

MacCannell, D. 1999. *The Tourist: A New Theory of the Leisure Class.* Berkeley, CA: University of California Press.

Massumi, B. 2002. *Parables for the Virtual: Movement, Affect, Sensation* .Durham, NC: Duke University Press.

Merleau-Ponty, M. 1968. *The Visible and the Invisible.* Evanston, IL: Northwestern University Press.

Obrador-Pons, P. 2003. Being-on-holiday: tourist dwelling, bodies and place. *Tourist Studies,* 3(1), 47–66.

Obrador-Pons, P. 2007. A haptic geography of the beach: naked bodies, vision and Touch. *Social & Cultural Geography,* 8(1), 123–41.

Obrador-Pons, P. 2009. Building castles in the sand: repositioning touch on the beach *Senses and Society,* 4(2), 195–210.

Obrador-Pons, P., Crang, M. and Travlou, P. 2009. *Cultures of Mass Tourism: Doing the Mediterranean in the Age of Banal Mobilities* (Aldershot, England: Ashgate).

Paterson, M. 2004. Caresses, excesses, intimacies and estrangements. *Angelaki,* 9(1), 165–77.

Paterson, M. 2006. Feel the presence: Technologies of touch and distance. *Environment and Planning D: Society and Space,* 24(5), 691–708.

Paterson, M. 2007. *The Senses of Touch: Haptics, Affects and Technologies.* Oxford: Berg.

Paterson, M. 2009. Haptic geographies: ethnography, haptic knowledges and sensuous dispositions. *Progress in Human Geography,* 33(6), 1–23.

Picart, C. 2002. Dancing through different worlds: An auto-ethnography of the interactive body and virtual emotions in ballroom dance. *Qualitative Enquiry,* 8(3), 348–61.

Pritchard, A. 2007. *Tourism and Gender: Embodiment, Sensuality and Experience.* London: CABI Publishing.

Radley, A. 1995. The elusory body and social constructionist theory. *Body & Society* 1(2), 3–23.

Saldanha, A.2002. Music tourism and factions of bodies in Goa. *Tourist Studies,* 2(1), 43–62.

Shields, R. 1991. *Places on the Margin: Alternative Geographies of Modernity.* London: Routledge.

Stranger, M. 1999. The aesthetics of risk: A study of surfing. *International Review for the Sociology of Sport,* 34(3), 265–76.

Straughan, E.R. 2010. 'Touched by water: The body in scuba diving. *Emotion, Space and Society*, doi:10.1016/j.emospa.2010.10.003.

Thrift, N. 1997. The still point: Resistance, expressive embodiment and dance, in *Geographies of Resistance*, edited by S. Pile and M. Keith. London: Routledge.

Thrift, N. and Dewsbury, J.D. 2000. Dead geographies and how to make them alive. *Environment and Planning D: Society and Space,* 18(4), 411–32.

Tuan, Y.F. 2005. The pleasures of touch, *The Book of Touch*, edited by in C. Classen. Oxford: Berg.

Urbain, J.D. 2003. *At the Beach.* Minneapolis, MN: University of Minnesota Press.

Urry, J. 1990. *The Tourist Gaze: Leisure and Travel in Contemporary Societies.* London: Sage.

Vasseleu, C. 1998. *Textures of Light: Vision and Touch in Irigaray, Levinas, and Merleau-Ponty*. London: Routledge.

Wylie, J. 2002. An essay on ascending Glastonbury Tor. *Geoforum,* 33(4), 441–54.

Chapter 3

Touching Space in Hurt and Healing: Exploring Experiences of Illness and Recovery Through Tactile Art

Amanda Bingley

Tactile and visual art in therapy is a powerful form of communicating feelings that are difficult to express verbally (Malchiodi 2003). Within art making, as art in therapy is termed, the sense of touch at the interface between skin surface and external tactile media plays a crucial role; connecting inner and outer worlds, holding and furnishing personal stories of physical and emotional experience. Touch and the haptic are thought to have this particular resonance in therapy because of specific pathways in the brain that link tactile/haptic perception with activated emotion and other related right sided brain function (Lusebrink 2004). The richness and therapeutic potential of the link between creative play and healing was noticed long before neuroscience confirmed these pathways in the brain. Early twentieth century psychoanalysts, notably the innovators of object relations theory Klein and Winnicott, and analysts who developed their ideas and those of Jung, were aware of the limitations of verbal expression in work with children in particular, but later with adults who had communication difficulties. They were quick to explore the wealth of non-verbal emotional expression through the world of play. Sand play, developed by Jungian therapists such as Kalff (1980) and Lowenfeld (1979), encouraged play with objects in a sand box to represent the child's inner imagined world. Klein, Anna Freud and Winnicott all independently developed observation of play with toys as an analytic tool[1]. In the decades following their pioneering work these concepts were developed in some instances to form distinct professions such as play therapy (Axline 1969) and art therapy (see Vick 2003). More generally a range of therapeutic artistic expression has evolved using visual and tactile art, music, drama and writing, referred to collectively as the 'creative' or 'expressive arts therapies' (Malchiodi 2003, Johnson 1999).

Art therapy has increasingly been offered by therapists working in private and community mental health services and during the last twenty years more routinely provided as part of hospice and charitable psycho-social support for people, and their carers, at different stages of treatment and palliation in life-threatening illness

---

1   Winnicott also developed the 'squiggle game', where the child's or adult's randomly drawn 'squiggle' served as a starting point for free association work by therapist and client.

(Wood 1998a and b). Extending the provision of art therapy beyond mental health services to those people with serious, physical conditions is in response to the particular and profound effect of illness on every aspect of the psyche. Whether acute or chronic, all illness results to some degree of feeling 'out of touch' with the familiar self. The moment of diagnosis with a life-threatening illness such as cancer is frequently described as a devastating disruption of oneself, a feeling of one's whole world falling apart – the continuity of self utterly and sometimes irretrievably left disconnected (Bingley et al 2006, Mathieson and Stam 1995). From the moment of diagnosis onwards people describe a constant struggle to regain or even locate their 'old selves' often felt to be more or less impossible during long periods of invasive medical treatments, however necessary (Barnard 1990, Thomas-Maclean 2004a). Öster et al (2007 and 2009) describe this experience as a loss of boundaries at a physical, psycho-social and emotional level. Their research with women during and following breast cancer treatment suggests that a creative activity, such as art making, supports individuals to repair disrupted boundaries and forge a re-connection with the self, a process that very often includes resolving feelings about childhood events and relationships.

An important aspect of art making as a therapeutic activity lies in the nature and function of touch in the physical process of exploring and expressing embodied experience through the medium of art materials. Tactile stimulation in therapeutic art making is, however, often overlooked and a gap remains in theoretical and empirical knowledge about its role, despite an increase in interdisciplinary literature, for example, in the sociology of health and illness, social anthropology, geography and art therapy, which theorises on tactile phenomena in the creative arts, therapeutic interventions, health, sensory perception in place and space, and technologies (Paterson 2007, Sibbett 2005). Touch and the tactile experience in therapeutic art making are the 'ground' from which two crucial therapeutic processes are facilitated. First, the physicality of working with tactile art materials like clay, sand and textiles acts as a highly effective 'mediation' or 'transitional object' between inner and outer self, which actively encourages the 'narration' of complex and often finely nuanced reflections on the illness experience. Second, due to the nature of the tactile as the primal and essential sensory mechanism, tactile art making methods enable a connection with the inner sense of 'authentic', 'true' self or personal 'idiom' (Bollas 1987, Mitchell 1993, Winnicott 1960); and a connection with a sense of the self in play (Winnicott 1971). As Wright (1991, 55) notes, the 'founding of the self' involves, what he describes as, '"good" bodily experiences with the mother' and that this is achieved through 'playful and loving interaction'[2].

This chapter examines some of these concepts, firstly around the role of the tactile in art techniques used in therapeutic and research contexts. I discuss

---

2   Although Wright's (2009) work is concerned with the role of vision in the development of self he suggests that the creation of self starts from the 'primary tactile' and then seeing what is touched.

why touch is relegated to the fringe of awareness in a complex and ambivalent relationship, even in therapeutic creative arts. The focus here is on the tactile element of touch, using the terms 'tactile' and 'touch' to describe and discuss the sensory experience of touching, with less emphasis on the collective term 'haptic'.

Secondly, I look at how touch in art making seems to facilitate the expression, resolution and emotional healing of the trauma of a serious illness experience, including where used in palliative care, by enabling a (re)connection with, what may be described as an 'authentic' sense of self. Lastly, building on previous research with tactile art, I theorise from an object relations perspective, that the tactile sense can be imagined as a 'micro-spatial' interface that holds, transmits and creates embodied past and present experience (Bingley 2002 and 2003, Bingley and Milligan 2007). In this way therapeutic art making allows the individual to engage with a fundamental phenomenon of self experience, integral to early development and subsequent adult life, where our relationship between and within inner self and the outer world (other) is mediated in an 'intermediate area of experience' via a 'transitional object' (Winnicott 1971), represented by the object of art making.

Touch Relegated to the Therapeutic Fringe

Touch is a complex of sensation pathways which includes the tactile (a term that refers specifically to the cutaneous sense of pressure), proprioception (perception of bodily position in space), kinaesthesia (bodily movement), and the cutaneous (skin sensations like temperature and pain) referred to collectively as the 'haptic' (Paterson 2007, ix). Many literatures explore the various aspects of the therapeutic qualities of being touched when someone is in need of medical or psychotherapeutic treatment, nursing, support and healing (Field 2001, Totton 2003), including a significant contribution from the development of specific 'therapeutic touch' techniques (Krieger 1979, Moore 2004). Other literatures debate the crucial importance of being touched as infants and children in order to initiate and promote essential brain development and emotional maturation (Montagu 1986). For example, the result of inadequate tactile stimulation in infancy and early childhood with little or no loving touch of stroking, cuddling and handling, and without the freedom to explore and play using touch, is so serious that our brain cannot develop properly; we are left emotionally and physically impaired and *in extremis* severe tactile neglect may end in death (Field 2001, Gerhardt 2004). To be able to touch other human beings or domestic pets, natural or made objects, to perceive varied textures, patterns and shapes, to differentiate temperature and pliancy is a crucial element of our sensory development and, literally, a touchstone throughout our adult lives (Field et al 1998).

There is, however, surprisingly little discussion about the sensory processes, specifically touch, involved in therapeutic art making. This lack in the literature is not immediately obvious, even to someone with a passing knowledge of using art as therapy. One might presume that the field of 'creative arts therapy' would be rich in debate about sensory mechanisms in art making. Creative arts therapy, as a profession, has a well established literature, covering the benefits and challenges of using 'art' therapeutically in myriad forms: visual, tactile, music, drama and writing (Malchiodi 2003). Authors in the field, though, are primarily involved in examining and applying psychoanalytic theories in relation to work with adults and children, and reporting on the use of creative arts in a variety of clinical, therapeutic and community settings including palliative care, psychotherapy, cancer care support and so on (Camic 2008, Staricoff 2004). With the call to develop more research in the profession (Gilroy 2006), interest is slowly starting to coalesce around the sensory, somatic mechanisms of therapeutic art. This is in tandem with the growing interest in the neuroscience behind the effects of creative arts and, for example, their use within the paradigm of cognitive-behaviour approaches to therapeutic interventions (Johnson 2009, Lusebrink 2004). Yet the process of touching, as the primary therapeutic *modus operandi* in tactile art therapies, remains relatively unexplored, even where tactile art is well established in supporting healing in life limiting illness during recovery, survivorship and at the end of life (Pratt and Wood 1998, Sibbett 2005). The fact that tactile experience is less visible in creative arts therapies echoes a cultural 'hierarchy' of senses that tends to prioritise the visual over the tactile (Paterson 2007, Smith 2007). What drives this ambivalence and why is touch relegated to the fringe of our awareness?

Touch is our primal sense (see Introduction, this volume, Montagu 1986, Streri 2003) and the subject of specific philosophical and scientific study, particularly since the nineteenth century. Integral to total sensory experience, we may underestimate or never bring to conscious awareness how reliant we are on tactility. The first sense to arise in embryo and the last to leave as we die, the conscious and unconscious relationship to touch and the tactile is virtually impossible to grasp. Gilman (1993) contends that it is the 'complex' and 'undifferentiated' nature of touch that make it the most difficult sense to study. He is concerned with the paradox of touch reflected in our socially constructed juxtaposition of medicalised versus sexualised touch and the ways we judge it, separating it as 'the erotic or painful' and as 'good touching or bad touching' (Gilman 1993, 199). Smith (2007, 94), in his work on Western sensory histories, offers a more mundane reason why touch has tended to be ignored, even 'slighted', historically: scholars have often found it hard to grasp and represent the sense of touch in a textual form, tending to prioritise visual imagery in historical accounts of events overlooking, sometimes, very rich tactile description in the written record. Indeed, as both Gilman and Smith note, sensory hierarchy in Western cultures is rarely discussed without revealing some ambivalence about touch. One of the only exceptions is when touch is the subject of scientific research and even this has, until recently, largely been the domain of enquiry into the experience of visually impaired or blind people who are particularly reliant on tactile stimuli.

In the last few decades technological advancements in neuroscience have led to more exacting research in our understanding of touch in biology, physiology and the psychology of touch in blindness including the mechanisms of learning to read Braille (Heller 1991, Millar 1997). Sensory development and modality is still not fully understood but is now thought to be cross-modal (also termed 'intermodal') (Shimojo and Shams 2001). This means that one sense does not develop independently from the others, rather from earliest infancy there is a continual 'modulation' between all five senses to create total sensory perception and interpretation of internal and external stimuli. Streri (2005) argues that cross-modal or intermodal development can be demonstrated in newborn infants. The pivotal sense in this process is the haptic, which Streri (2003, 51) states is the most 'primitive sense', demonstrable in 'the first few weeks of foetal life'. Tactile perception at the cutaneous surface is also found to be exquisitely sensitive, particularly over the surface of the hands and fingertips, an area of the skin rich in nerve endings. Indeed, throughout our lives, we can differentiate, just by touch, between micro millimetres of thickness in textured materials (Gentaz and Hatwell 2003, Hollins et al 1999). Given this capacity and the concentrating of tactile perception in the hands and fingertips, we can appreciate the ways touch is integral to our sensory development and utility. As Streri (2005, 326) affirms:

> Hands are a complex system that involves two functions: a perceptual function ("knowing") and an instrumental function ("doing") ... The hands are also the motor organs used to reach, hold, transform and transport objects in our everyday life. This second function is specific to the manual system and gives it a unique and original characteristic among the senses. Thus, we can create events with our hands, even though these productions are often limited.

Therefore, from infancy we engage in what could be termed a continual 'haptic exchange': touching our own body, touching carers and objects in our immediate environment with hands and fingertips, mouths and lips; in active physical movements, learning to roll over, crawl and then walk. Although learning about our internal, embodied and external world may require intermodal sensory interaction, we start with the tactile.

Studies of intermodality using comparative sensory perception between sighted versus visually impaired or blind individuals shows there is a particularly strong relationship between vision and touch[3]. In a study by Heaps and Handel (1999) participants shown two-dimensional images of textured surfaces have no difficulty accurately grouping these into rougher or smoother surfaces, using only visual cues learnt intermodally in conjunction with tactile stimuli. Conversely,

---

3   First described by seventeenth century philosopher William Molyneux, there was much philosophical debate, later known as 'Molyneux's question', about whether the senses developed and operated independently or intermodally, and how this process was affected by blindness (Paterson 2007). Although congenital or acquired visual impairment pose constraints on total sensory organisation and perception, individuals are still able to adapt and acquire knowledge of the external world relying on the 'ground' of the haptic and working intermodally with aural, olfactory and gustatory perception.

congenitally blind individuals, whose vision has eventually been restored are unable to accurately identify an object visually they have previously only known through touch, until they are able to explore the object together with vision and touch. These studies do, however, emphasise the essential nature of the sensory 'ground' that tactile experience provides and through which we are able to 'make sense' of perceived external stimuli and internal, embodied (somatic) phenomena.

In the course of childhood development and into adulthood we can and often do, if we are not visually impaired, let the tactile, indeed the full haptic, fade into the background of our awareness. Vision, and the safe distance it affords, allows us such a rapid assessment of our surroundings at any given time that as a result we may imagine and consciously promote vision as the key sense through which taste, smell, sound and touch are interpreted. Touch though, as suggested above, adds a crucial spatial and experiential dimension to the assessment of internal sensation and external stimuli. While we may speak of the eyes as the window to the soul, tactile exchange is the 'ground' from which our visual connection and interpretation arises and the 'ground' through which we reference the other senses of taste, smell, and sound. This idea of a ground or baseline of primal sensory connection with the earliest elements of the foundations of self echoes Streri's (2003) earlier assertion of the primitive (and primal) place of the haptic from earliest infant experience. The ground of tactile exchange is where, we could say, the '*macro* space' of our external world interfaces with our inner at what may be conceived as a tactile '*micro* space' on the cutaneous boundary of the physical body. This interface corresponds with Winnicott's (1971) 'intermediate area of experience' or 'potential space' where we mediate inner and outer experience and relationships between our 'self' and 'not self' or 'other'. In other words the tactile 'micro-space' is a crucially important element of mediating our experience of and relationship with self. These concepts are important in understanding the potential therapeutic role of tactile perception in art making, as I shall discuss below. First, I shall define in greater depth the concepts of 'self' and 'not-self' or 'other'.

Reconnecting with an Authentic Sense of Self in Hurt and Healing

Focused, aware touching in the form of art making is a creative opportunity to reconnect us to our 'primal ground'. Being in touch, though, does more than connect us with our creativity. The unique primacy and intimacy of touch connects us with a subjective sense of self. The idea of self, however, as an identifiable, defined subjectivity is notoriously elusive. As Pile and Thrift (1995, 9) argue, the self, identity and subjectivity are spatially integral with the social thus the self is always 'multiple, moving, changing'; in other words, the self is always under construction as a social identity and as such can never be defined. The subjective experience of self is no less elusive and mercurial. As fast as we create and recreate a 'trope' of experience, memories, thoughts and ideas that emphasise the continuity of a sense of self as a stable, immutable and authentic identity, that moment will be gone risking having to face what Laclau (1994, 3, cited in Pile and Thrift 1995, 9) has described as the lack 'at the root of identity'. Hence, subjectively the sense of self, which must, by default, be experienced as truly 'me' having some authenticity, must be continuously invoked to avoid an imagined catastrophic collapse of identity – a chaos of 'no self' into all that is outside of the physical and psychic boundaries known as 'not-self' or 'other'. The idea of reconnecting with an 'authentic sense of self', requires moving into what Taylor (1989, 175–6, cited in Pile and Thrift 1995, 8), describes as a 'reflexive stance', in which '[w]e have to turn inward and become aware of our own activity and of the processes that form us'. From an object relations theory perspective, despite the individual's capacity to be reflexive, the self is fundamentally 'unknowable' regardless of any subjective experience of feeling the self as 'real' or 'conversely' ... 'false, incoherent, fragmented' (Bingley 2002, 35). In this model the self is imagined as a semi-submerged consciousness, located within our physical, sensory boundaries that flows back and forth from our unconscious in waking and dream life (see Bollas 1987). Despite the elusive nature of self, we will always seek a feeling of authenticity: we are bound to in order to maintain a functioning personhood. An inner sense of self as 'authentic' or 'true' (Mitchell 1993) versus a sense of 'false self' was originally a focus of Winnicott's theorising in his exploration of the differing states that people perceive themselves in mental distress versus in good mental health. Winnicott (1960, 148) suggested that re-finding and living from our sense of 'true self' was part of the successful outcome of analysis:

> Only the True Self can be creative, and only the True Self can feel real ... [In contrast] the existence of a False Self results in feeling unreal or a sense of futility ... The True Self comes from the aliveness of the body tissues and the working of the body functions.

Contemporary psychoanalyst Christopher Bollas (1987), in developing Winnicottian theory, further modulates this version of self by focusing on the concept of the 'aliveness' of the 'true self'. He sees the self not as a comparative binary of 'false' or 'true' but rather an experience of authenticity arising from our inner self or 'idiom'. As Bollas (1987, 9) affirms: 'it is important to stress how this core self is the unique presence of being that each of us is; the idiom of our personality'. More recently, he has refined his concept of self as idiom, cogently describing the self, or our reference point of 'me', as an 'inner constellation', that in effect is virtually unknowable and indescribable in totality. As he reflects:

> when we think "me" without reference to any other term, we evoke a dense inner constellation, a psychic texture, existing not in the imaginary, although it yields derivatives there, but in the real, an area that can be experienced but cannot be represented in itself ... The me can be conceptually identified and its material discussed. It is composed of memories (including the history of desire), and

these constitute the cumulative psychic outcome of idiom's theories and their
enacted deployment in a life's experience. (Bollas 1995, 152)

In this psychoanalytic model, therefore, the self can be conceptualised as a
'constellation', or collage, that as a whole and in each part, has a unique story,
collection of memories and 'psychic texture' expressing an individual's own
'peculiar essence or a personal idiom'. Realising and connecting with our idiom
and then expressing this as a fully conscious, recognisable total experience is,
Bollas suggests, an impossibility, not least because much of our self lies beyond
our conscious thought in the realms of unconscious dreaming. As such we can
never grasp the whole, only the dream shadows of our self as idiom. Being and
living our idiom is a process, a continual becoming of self within embodied
sensory and somatic experience.

As Wright (2009) emphasises the primal elements of a sense of self are
embedded in the primary relationship between infant and the 'holding' by the
'good enough' mother. Thence, development and maturation of embodied self
continues in the 'intermediate area of experience' or the Winnicottian 'transitional
space' between perceived self and other. Thereafter, to maintain continuity of
self, feeling 'in touch' with our 'sense of self' requires sensory (and cognitive)
revisiting/ (re) connecting with embodied elements of the primary relationship.
Creating art, Wright (2009) contends, whether tactile or visual, has an important
function as self-expression, both reiterating and communicating the continuity and
continual becoming of self as real and authentic. But what is perceived to happen
to the continuity of our embodied self during illness?

In states where the self has been disrupted by some traumatic event or process,
as in serious illness and invasive treatments, the self is experienced as incoherent
and fragmented, described by the sufferer as feeling 'not my self' and less able to
function at every level, physically, mentally or emotionally (see Barnard 1990).
The disruptive effects of serious illness on our embodied self start from the point
where symptoms impinge on our everyday life, but many people relating their
illness narratives report diagnosis as the single biggest impact on their self-identity
and on their personal and social relationships (Bingley et al 2006). Subsequent
invasive tests, treatments and surgery are experienced at a visceral, primal level
as hurts, however well-meaning. As Gilman (1993) notes, this is the paradox of
medicalised touch: it hurts. At the most primitive level of subjective embodiment
we perceive painful procedures as the body undergoing, what could be termed
as, a micro-spatial invasion within the macro-spaces of the medical environment.
The resulting disruption, as Öster et al (2009) describe in their work with women
in recovery from breast cancer treatment, occurs at both the actual physical and
imagined boundaries of self.

In terms of sensory perception the disruption of self occurs within the total
haptic experience, internally and externally. The greater the disruption, the greater
the impact on the perceived sense of self until there is felt to be a threat to the very
authenticity of self. Hence, the individual is in a constant struggle to re-negotiate

the 'ground' of the embodied self. If the illness results in a permanent and/or life-limiting state of chronic ill-health or disability, the individual has to find a way, somehow, however fleetingly, to re-connect with their inner authentic self, to 'feel themselves' again, or risk what is variously described as a state of dislocated, alienated self, in a narrative of chaos and despair, 'out of touch' with self (Frank 1995, Thomas-MacLean 2004b). However, inasmuch as we are hurt by medicalised touch, we can equally heal by using the therapeutic qualities of touch. For instance, in therapeutic art making, at one level we seek to repair the self hurt by 'bad' touch by using 'good' touch to heal the self. At this most primal level the organism as embodied self may perceive and interpret embodied experience at an unconscious and at times conscious level. Bringing somatic responses and emotional feelings into conscious awareness through the art making is a way of facilitating a re-negotiation and repair of boundaries. Wright (2009) expands on this idea of art making as way of reconnecting with the embodied self by invoking the primal relationship. He argues that in creating an expression of self in the art work we also communicate elements of self to the 'not-self'/other and this affirms the subjective sense of authentic self as real. Sensory experience is integral to the primal and subsequent sense of embodied self, and as I argued above, the tactile and haptic are the crucial 'ground' from which elements of self are continuously (re)created.

In attempting to recover from and resolve the traumatic effects of life-limiting illness, then, we can consciously make use of touch in art making in two important ways. Firstly, as I describe above, touch in art making is a way of literally getting 'back in touch' with a sense of the familiar self; to re-establish the primacy of a sense of 'authenticity' of self; those known limits of self that are felt to be lost in the disruption of serious illness and invasive medical treatments. Secondly, the act of art making is reported as facilitating the healing of and reconciliation with past traumas. Past hurts can surface and be re-activated with a particular urgency when we face life threatening illness. In the course of healing, though, there is often an openness to be reconciled with those events whether as a resolution at the end of our life, or as part of our recovery and survivorship. For example, in palliative care art therapy, the urgency to resolve past hurts and reassert a sense of self identity is well-documented (Pratt and Wood 1998, Sibbett 2005). A patient may want to engage in therapeutic art making, even in the last stages of illness when they are barely able to do any of the physical activity, but where they can still touch and feel the texture of an image or artefact. Stone Matho (2005), an art therapist working in supportive and palliative care in France, relates how one patient had benefited from art sessions over many months as a way to resolve feelings from early family relationships thus allowing reconciliation with her father. This patient was so determined to complete her process that she insisted to the end of her life to continue art making actively holding and handling art materials, with Stone Matho eventually taking the materials to the patient's bedside. Although, Stone Matho's description remains focused on the patient's process art as therapy, inevitably touching and working with the art materials was integral to the process.

Touching in Therapeutic Art Making

The German art therapist Uwe Herrmann (1995) reports that when he started work in the early 1990s at a residential state school for blind children, he discovered that the pupils had been denied any art making in their education, on the basis that it was visual and thus has no place in a school for the blind. He noted that 'to give not only clay but also paint to blind or partially sighted students was still regarded as a revolutionary act by quite a few staff' (Herrmann, 1995, 229). Yet he was able to counter this situation by instituting art sessions and demonstrating that tactile art was a valuable and therapeutic medium for blind children, some of whom were deeply traumatised refugees from the conflicts in former Yugoslavia. The sessions helped them to express and process their feelings about often harrowing experiences. Herrmann's (1995) paper is, however, a very rare example by an art therapist *specifically* describing the importance of the *tactile* qualities of art making. What is less unusual is that he describes the benefits of tactile art in the context of working with the blind and visually impaired.

In psychoanalytic and psychotherapeutic settings there is great sensitivity towards the subject of touch and touching with some heart-searching scrutiny and debate, much of it concerned with the importance of maintaining integrity and proper conduct of practitioners, whether their work is 'verbal' or 'body' psychotherapy (Totton 2003). Creative arts therapists regard touch and touching, where appropriate, as a natural and integral part of the therapeutic experience, despite the apparent lack of interest in recognising and exploring the tactile element of art making. The few exceptions in the literature include the work of neurobiologist Lusebrink (2004) who has examined which areas of the brain are stimulated during art therapy activities. From an art therapist perspective Sibbett (2005) also describes the importance of 'liminal' spaces in art making, meaning the spaces at the edge or interface of experience, actual and imagined, between the individual and their art. While Wood (1998b) mentions the physicality in art therapy, describing this as 'bodyliness', she also refers to an observation by Erskine and Judd (1994) who, she notes, maintain that art therapy provides an opportunity to feel a 'unity' of 'body and mind'; as Wood (1998b, 34–35) further explains:

> This interrelationship of body and mind is extremely difficult to experience. Something of this unity can however be glimpsed in artwork produced in therapy. The physicality of art materials along with their symbolic capacity enables wordless layers of experience to be rendered in concrete form, with the resonances between the two being regarded as therapeutically powerful.

Erskine and Judd, from Wood's interpretation, come closer than other art therapy literatures to a tacit acknowledgement of the importance of the tactile connection in art making, by which we can communicate and make sense of aspects of self experience. In the final section of the chapter, I explore the use of this interface in tactile art and how these principles may be involved when seeking insight or resolution in illness.

Touching the Transitional Object: Processes of Self-expression through Tactile Art

Figure 3.1   Touching spaces in sandplay

Source: author photograph

In therapeutic art making touch, I suggest, holds an often unacknowledged and hidden primacy in the interaction between an individual and the different materials used in sessions. Art materials may include paints, paper, sand, clay, textiles and metals, and natural materials such as wood and stone. Activities can include holding a brush or pencil, or using fingers and hands to apply paint or create a two-dimensional image or collage; fashioning artefacts and incorporating natural objects into a three dimensional model. Even in painting the very act of picking up a brush or pencil connects us instantly with the tactile connection or conduit, 'grounding' us between and within inner self and outer other (paper or art medium). In art making workshops I have found that highly tactile media such as sand, clay or textiles, are particularly effective as expressive media. Sandplay, for example, a non-verbal technique often used in art therapy, involves literally playing with wet and dry sand as a medium with which to explore and express thoughts and feelings difficult to access or communicate verbally (Figure 3.1). These feelings maybe consciously associated with some specific situation or experience or the participant or client may be feeling some undefined mental and emotional state at the fringes of everyday consciousness, which when accessed and expressed via the non-verbal art making can facilitate insight and / or resolution (Wood 1998a).

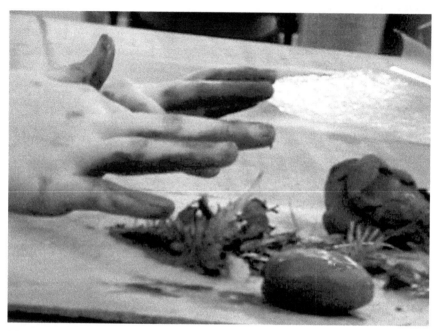

Figure 3.2   Fully in touch with clay

Source: author photograph

Clay is also used extensively, and working with a single medium of sand or clay seems to help focus participants on the therapeutic, mediating expressive process rather than striving to create art[4]. Touching and working with wet or dry sand and clay is a strong, tactile, indeed fully haptic, stimulus of hands and fingers using a range of delicate and rough muscular movements that necessarily involve the whole body; this effectively ensures a total sensory engagement, concentration and focus with the task[5] (Figure 3.2). For example, within minutes of starting to touch the sand, clay or textiles participants will use a variety of interactions:

---

4   Theories and practice of sand play are extensively discussed elsewhere (see Bingley, 2003; and Bingley and Milligan 2007 for examples of using this technique in geographic fieldwork).

5   In comparison tapping a digital keyboard, for example, does not create the same intense tactile stimulus, due to the small, uniform surface area involved and the limited potential for creative tactile interaction with the medium of a digital keyboard.

stroking, smoothing, patting or hitting, moulding and fashioning shapes, kneading and sculpting sand or clay. They report feeling almost instantly and intensely absorbed in the activity. For example, research participants in an art making workshop exploring tactile experience with sand, clay, wood and textiles describe working with the different materials in exacting detail with their eyes shut and focused on touch:

> an amazing bit of velvet here, really so soft, incredible' ... 'soft and warm [velvet] and the clay is cold and clammy'...'the contrast [between clay and velvet] is quite a shock really, your hand on the clay, although I loved it, it was rougher (Bingley 2002, 4498A Notes).

In another workshop sand, in particular, was found to induce:

> a dreamy state ...' and that it was possible to 'get completely immersed in it [the sandplay]'... [reminding the participants] 'of being very young' ... 'I really love the texture. I love wet sand a lot more than dry sand, 'cos wet sand it doesn't go straight through your fingers, you know, you can like hold it there and, and everything and I don't know, I was just really enjoying myself. I just sat there and thought 'wow! I'm like a kid again. ... and it's really good. (Bingley and Milligan 2004, QE5 workshop)

The art making, in effect, becomes the psychoanalytic 'object'[6] and this as Wright (2009) argues represents the Winnicottian 'transitional object' which mediates between self and other; a neutral, unconditional object that can be made use of freely by the unconscious self for projection or transference into the conscious self, allowing resolution. Similar to Wright (2009), I suggest that this kind of art making can be a medium by which various primal needs are re-enacted stimulated by touch and touching, including the earliest embodied memories of being soothed and stroked either by our own touch or the touch of our primary carer(s), as well as early disillusionments and traumatic physical, tactile experiences. Thus, either consciously or unconsciously, the object (represented by sand, clay or other art materials) is used to attempt to reconnect with the primal 'ground' and to resolve both the current and any early trauma (whether or not we bring this part of the process to our conscious awareness). In effect, the apparent simplicity of 'just playing' with sand or clay underestimates the power of the tactile medium as transitional object to act as mediator between self and other.

In a situation where the self is disrupted and the boundaries of self traumatised by illness, art making is found to support an individual to resolve the disruption, repair and 'strengthen' the boundaries (Öster et al 2007, 277). Although, as noted

---

6   The psychoanalytic 'object' is defined by Bollas (2009, 15) as any 'thing' (object) that 'can stand in for 'parts' of the self in relation to his or her mental objects, usually differing forms of representation of other people'.

by Öster et al (2007) exactly how art making facilitates this process is unknown, there is an extensive, and impressive, literature giving case by case examples of the therapeutic potential and beneficial results of art making (Jones 2005, Thomas-MacLean 2004a, Malchiodi 2003). I suggest that these therapeutic qualities of art making as a way of supporting recovery or resolution in illness is, in part, due to the expression of self through the ground of the tactile and consequent opportunity to reconnect with a sense of the 'old self'; that is the sense of 'authentic' or 'true' self. Recognising this may help us make sense of why embodied processes of art making appear so effective in facilitating a sense of healing from the traumatic disruptions of illness and resolution of difficult emotions.

Tactile art making will almost invariably at some point, as I reported above, bring out childhood memories. This phenomenon is well-documented in creative arts therapy literature (Malchiodi 2003). The individual is emboldened to express and explore these past experiences and embodied thoughts as they arise through the art making process, and to reconnect with the 'playful' self. Through this re-connection with the primal tactile ground and various elements of self, we re-assert our sense of 'authenticity', (re)gaining a sense of inner authority and control, and thus able to re-negotiate current and future self/other relationships (Bollas 1987). In this way, working with the art object can be used to heal the disrupted narratives of self; a process that engages with our creative capacity to communicate and resolve powerful emotions (Wood 1998a).

Another important aspect of the therapeutic art making process is that the sessions are conducted in a 'safe space' where the process is 'held' and acknowledged within a 'facilitating environment'; that space between an individual art-maker and the therapist as witness (Marxen 2009, Winnicott 1971). Marxen (2009, 133), describing art therapy with traumatised young people, emphasises that the 'safe empathetic atmosphere' of the art therapy space allows the individual to safely express aggressive or difficult feelings through symbolism of the art making. Thus, the external space I describe as the '*macro*-space' of the art making environment acts to contain and allow some element of the individual controlling the expression of potentially overwhelming feelings that flow through *micro*-spaces of touching. At least some of this powerful process of art making and the connection with self is only possible in the context of a safe place where the art making can be 'held'. Creative art whether used in therapy or in a therapy research setting, is usually conducted in a space and time specifically set up for this purpose, facilitated by a trained psychotherapist who aims to 'hold' the therapeutic space enabling the participants (Winnicott 1971). In this context the therapist witnesses and supports the individual's process. In my own research with tactile methodologies and in my therapeutic experience as a psychotherapist, this is an essential aspect of the process: the macro-space of the art making needs to be adequately held by the facilitator /therapist, in order that the individual can feel safe enough to connect with self and other via the micro-space of touching the art object. The importance of the 'holding', in Winnicottian terms, is that the therapist protects the macro-space in the same way as 'good enough mothering' protects the child's early play space.

Conclusion

Art therapists and the individuals with whom they work have documented the benefits of using creative arts as an approach to healing, suggesting that art making can be a sensitive and effective medium for exploration of thoughts and feelings that are difficult to access and verbalise. Touch is an integral, but often overlooked aspect of art making and the tactile element of art making is a key factor in the therapeutic process. This may be due in part, I suggest, to the primacy of the tactile and haptic 'ground' in our earliest development of a sense of self and identity. The effect of connecting with our creativity via this tactile 'ground' facilitates a re-negotiation with a sense of self that is felt to be 'authentic' or 'true', and this process can be profoundly healing. Whilst the 'self' is acknowledged to be essentially unknowable, subjectively there is a need to maintain a sense of self and identity experienced broadly as feeling 'authentic' or 'true'. The experience of serious and life-threatening illness, fraught with long, often traumatic treatments and the possible end of life, is reported as a profound disruption to the sense and boundaries of self; physically, mentally and emotionally. This can result in a feeling of the self as 'fragmented' or 'false'; feeling 'out of touch' with 'oneself', with past pain and childhood memories often reactivated. Art therapy is found to be a powerful and effective means to seek some resolution and re-establish a sense of 'true' self.

The tactile sense at the physical boundary of hands and fingers is, I propose, a 'micro-space' between inner subjective experience and the macro-space of the external other. Art making using highly tactile media such as sand, clay and textiles can facilitate communication and expression via these micro-spaces of tactile perception between inner thoughts and feelings that are difficult to access and verbalise. An important aspect of the process is having a supportive witness to this expression, a role usually fulfilled by a trained therapist. The tactile element of therapeutic art making is considered to be one key to understanding the process by which creative arts can facilitate insight and emotional resolution for those suffering serious and life threatening illness.

References

Axline, V.M. 1969. *Play Therapy*. New York: Balantine.
Barnard, D. 1990. Healing the damaged self: identity, intimacy, and meaning in the lives of the chronically ill. *Perspectives in Biology and Medicine*, 33, 535–46.
Bingley, A.F. 2002. *The Influence of Gender on the Perception of Local Landscape*. University of Edinburgh: Unpublished PhD Thesis.

Bingley, A.F. 2003. In here and out there: sensations between self and landscape. *Social and Cultural Geography,* 4(3), 329–45.

Bingley, A.F. and Milligan, C. 2004. *Climbing Trees and Building Dens* (Unpublished research fieldnotes, Lancaster University)

Bingley, A.F. and Milligan, C. 2007. Sandplay, clay and sticks: multi-sensory research methods to explore the long-term influence of childhood play experience on mental well-being. *Children's Geographies,* 5(3), 283–96.

Bingley, A.F., McDermott, E., Thomas, C., Payne, S.A., Seymour, J.E. and Clark, D. 2006. Making sense of dying: a review of narratives written since 1950 by people facing death from cancer and other diseases. *Palliative Medicine,* 20(3), 183–95.

Bollas, C. 1987. *Forces of Destiny. Psychoanalysis and Human Idiom.* London: Free Association Books.

Bollas, C. 1995. *Cracking Up. The Work of Unconscious Experience.* London: Routledge.

Bollas, C. 2009. *The Evocative Object World.* London: Routledge.

Camic, P.M. 2008. Playing in the mud: health psychology, the arts and creative approaches to health care. *Journal of Health Psychology,* 13(2), 287–98.

Erskine, A. and Judd, D. 1994. *The Imaginative Body: Psychodynamic Therapy in Health Care.* London: Whurr.

Field, T. 2001. *Touch.* Cambridge, MA: MIT Press.

Field, T., Fernandez-Rief, M., Quintino M, Schanberg, S. and Kuhn, C. 1998. Massage therapy – elder retired volunteers benefit from giving massage therapy to infants. *Journal of Applied Gerontology,* 17, 229–39.

Frank, A.W. 1995. *The Wounded Storyteller: Body, Illness, and Ethics.* Chicago: University of Chicago Press.

Gentaz, E. and Hatwell, Y. 2003. Haptic processing of spatial and material object properties, in *Touching for Knowing,* edited by Y. Hatwell, A. Streri and E. Gentaz. Amsterdam: John Benjamin), 123–59.

Gerhardt, S. 2004. *Why Love Matters. How Affection Shapes a Baby's Brain.* London: Routledge.

Gilman, S. 1993. Touch, sexuality and disease, in *Medicine and the Five Senses,* edited by W.F. Bynum and R. Porter. Cambridge: Cambridge University Press, 198–224.

Gilroy, A. 2006. *Art Therapy, Research and Evidence-based Practice.* London: Sage.

Heaps, C. and Handel, S. 1999. Similarity and features of natural textures. *Journal of Experiment Psychology: Human Perception and Performance,* 25(2), 299–320.

Heller, M.A. 1991. Haptic perception in blind people, in *The Psychology of Touch,* edited by M.A. Morton and W. Schiff. Hillsdale NJ: Lawrence Erlbaum.

Herrmann, U. 1995. A Trojan horse of clay: art in a residential school for the blind. *The Arts in Psychotherapy,* 22(3), 229–34.

Hollins, M., Faldowski, R., Rao, S. and Young, F. 1999. Perceptual dimensions of tactile surface texture: a multidimensional scaling analysis. *Perception & Psychophysics,* 54(6), 697–705.

Johnson, D.R. 1999. *Essays on the Creative Arts Therapies.* Springfield, IL: Charles C Thomas.

Johnson, D.R. 2009. Commentary: examining underlying paradigms in the creative arts therapies of trauma. *The Arts in Psychotherapy,* 36, 114–20.

Jones, P. 2005. *The Arts Therapies: A Revolution in Healthcare.* Hove: Brunner-Routledge.

Kalff, D.M. 1980. *Sandplay.* Boston, MA: Sigo Press.

Krieger, D. 1979. *The Therapeutic Touch.* New York: Fireside.

Laclau, E. 1994. Introduction, in *The Making of Political Identities,* edited by E. Laclau. London: Verso, 1–8.

Lowenfeld, M. 1979. *The World Technique.* London: Allen & Unwin.

Lusebrink, V.B. 2004. Art therapy and the brain: an attempt to understand the underlying processes of art expression in therapy. *Art Therapy: Journal of the American Art Therapy Association,* 21(3), 125–35.

Malchiodi, C.A. 2003. Using art with medical support groups, in *Handbook of Art Therapy,* edited by C.A. Malchiodi. New York: Guilford Press, 351–61.

Marxen, E. 2009. Therapeutic thinking in contemporary art. Or psychotherapy in the arts. *The Arts in Psychotherapy,* 36, 131–39.

Mathieson, C.M. and Stam, H.J. 1995. Renegotiating identity: cancer narratives. *Sociology of Health & Illness,* 17(3), 283–306.

Millar, S. 1997. *Reading by Touch.* London: Routledge.

Mitchell, S.A. 1993 *Hope and Dread in Psychoanalysis* .New York: Basic Books.

Montagu, A. 1986. *Touching: The Human Significance of Skin, Third Edition.* London: Harper Collins.

Moore, T. 2004. *Annotated Bibliography of Published Therapeutic Touch Research: 1975 to July 2004.* Toronto: Therapeutic Touch Network of Ontario.

Öster, I., Magnusson, E., Thyme, K.E., Lindh, J. and Åström, S. 2007. Art therapy for women with breast cancer: the therapeutic consequences of boundary strengthening. *The Arts in Psychotherapy,* 34, 277–88.

Öster, I., Åström, S. Lindh, J. and Magnusson, E. 2009. Women with breast cancer and gendered limits and boundaries: art therapy as a 'safe space' for enacting alternative subject positions. *The Arts in Psychotherapy,* 36, 29–38.

Paterson, M. 2007. *The Senses of Touch: Haptics, Affects and Technologies.* Oxford: Berg.

Pile, S. and Thrift, N. 1995. Introduction, in *Mapping the Subject: Geographies of Cultural Transformation,* edited by S.Pile and N. Thrift. London: Routledge), 1–12.

Pratt, M. and Wood, M.J.M. 1998. *Art Therapy in Palliative Care.* London: Routledge.

Shimojo, S. and Shams, L. 2001. Sensory modalities are not separate modalities: plasticity and interactions. *Current Opinion in Neurobiology,* 11, 505–09.

Sibbett C. 2005. Liminal embodiment: embodied and sensory experience in cancer care and art therapy, in *Art Therapy and Cancer Care*, edited by D. Waller and C. Sibbett. Maidenhead, England: Open University Press, 50–81.

Smith, M.M. 2007. *Sensory History*. Oxford: Berg

Stone Matho, E. 2005. A woman with breast cancer in art therapy, in *Art therapy and Cancer Care*, edited by D. Waller and C. Sibbett. Maidenhead, England: Open University Press, 102–18.

Staricoff, R. 2004. *Arts in Health: a review of the medical literature*. Arts Council England, <www.artscouncil.org.uk/documents/publications/phpc0eMaS.pdf>.

Streri, A. 2003. Manual exploration and haptic perception in infants, in *Touching for Knowing*, edited by Y. Hatwell, A. Streri and E. Gentaz. Amsterdam: John Benjamin, 51–66.

Streri, A. 2005. Touching for knowing in infancy: the development of manual abilities in very young infants. *European Journal of Developmental Psychology,* 2(4), 325–43.

Taylor, C. 1989. *Sources of the Self: The Making of Modern Identity*. Cambridge: Cambridge University Press.

Thomas-MacLean, R. 2004a. Memories of treatment: the immediacy of breast cancer. *Qualitative Health Research,* 14(5), 628–43.

Thomas-MacLean, R. 2004b. Understanding breast stories via Frank's narrative types. *Social Science & Medicine,* 58, 1647–57.

Totton, N. 2003. *Body Psychotherapy: An Introduction*. Maidenhead, England: Open University Press.

Vick, R.M. 2003. A brief history of art therapy, in *Handbook of Art Therapy*, edited by C.A. Malchiodi. New York: Guilford Press, 5–15.

Wood, M.J.M. 1998a. What is art therapy?, in *Art Therapy in Palliative Care,* edited by M. Pratt and M.J.M. Wood. London: Routledge, 1–11.

Wood, M.J.M. 1998b. Art therapy in palliative care, in *Art Therapy in Palliative Care*, edited by M. Pratt and M.J.M. Wood. London: Routledge, 26–37.

Winnicott, D.W. 1960. Ego distortion in terms of true and false self, in *The Maturational Processes and the Facilitating Environment,* by D.W. Winnicott. London: Karnac Books, 37–55.

Winnicott, D.W. 1971. *Playing and Reality*. Harmondsworth, England: Penguin.

Wright, K. 1991. *Vision and Separation: Between Mother and Baby*. London: Free Association Books.

Wright, K. 2009. *Mirroring and Attunement: Self-realisation in Psychoanalysis and Art*. London: Routledge.

Chapter 4
Facing Touch in the Beauty Salon: Corporeal Anxiety

Elizabeth R. Straughan

Introduction

This chapter attends to the sense of touch as it is experienced and deployed within the beauty salon, a site that is exemplary of the 'body industry'. Comprised of diverse sites, practices and expertises, all promoting the 'virtues of a healthy, risk averting body' (Peterson and Lupton 1996, 24), the beauty industry is underwritten by a philosophical approach to modern medicine that considers the term 'health' to mean not only the 'absence of disease and infirmity', but also 'a state of complete physical, social and mental well-being' (Kearns 1993, 142, citing the World Health Organisation). What renders the beauty salon distinctive, however, is its traditionally 'feminised' approach to the aesthetic problematisation and conditioning of the surface of the body, and in particular the skin of the face, as well as the promulgation of treatments that, under the expert hands of trained personnel, promise to ameliorate aesthetic concerns. Indeed, within the beauty salon the application of products and their promised functionality goes hand in hand with the provision of a therapeutic or relaxing experience (Straughan 2010).

In focusing on touch in this context, then, I draw particular attention to a 'working with skin,' by which is meant not only the 'passive' surface of the face, as well as the cutaneous and subcutaneous sensitivity of the hand, but also the emotional mobilisation of the psyche or self. That is, the feel of receiving and giving facial treatments draws out touch as a mode of corporeal manipulation, as well as the role touch plays in establishing client well-being. A key inspiration for this discussion is the work of Luce Irigaray (1993) and her concept of 'porosity,' by which she refers to the skin's capacity for vulnerability, and the manner in which entities such as the beauty salon takes advantage of the bodies continual 'a-morphe' or 'gestations.' Accordingly, a crucial grounding for this comes from my own experience of receiving treatments, such that the site of geographic enquiry becomes consolidated onto the skin of my face. I follow Irigaray (1992, 51) in considering '[t]he open horizon of my body. A living, moving border. Changed through contact with your body [and the body of the beautician, I would add]'. It is in attending to the corporeal fluidity of Irigaray's work, and the manner in which touch invokes the fallibility and fragmentation of the body that, I go on to suggest, the beauty salon can be understood as a site wherein we can find what Katherine N. Hayles (1997) has termed 'corporeal anxiety' at work.

Irigaray and Touch

Irigaray (1993) suggests that touch is premised upon the permeability of bodily boundaries, exemplified through a consideration of a feminine, corporeal materiality. She produces a particular imaginary of the feminine lips and the tangible passage of mucus from inside to outside. Drawing on the ability of lips to touch one another such that the *gap between* advocated by a masculinist philosophy becomes problematised, Irigaray highlights the openness of entities to each other and the world. In doing so, Irigaray's focus tends to the general rather than the specific, to be sure, an attendance often critiqued by feminist thinkers (such as Rose 2003). Yet, it is the openness suggested by Irigaray's work, and the facilitation of passage that her focus on touch draws out, which highlights the 'temporality' of Irigaray's philosophy, a temporality that works to unsettle such critiques.

According to Ziarek (1998, 67), for example, 'this inaugural temporality has to be linked with the becoming of the body'. The female imaginary of touch, mucus, and threshold offered by Irigaray is predicated upon a sense of movement, 'ceaselessly reshaping this incarnation' or, rather, the body (Irigaray 1993, 67). Touch signifies porosity only in its ability to, 'metabolise itself in the constitution of time' (1993, 191), for it is only in attendance to a duration of the body as 'a *morphe* in continual gestation', that the fluidity between inside and outside is recognisable (1993, 193). And so, while Irigaray's philosophy is focused on the biology of the body, underscoring this is an acknowledgement of its activity and changeability. Importantly, this changeability is not centred on a corporeal materiality alone, it also suggests an openness to culture, for 'bodies ingest culture to make themselves, and so culture becomes corporeal' (Rose 2003, 56).

As a number of geographers have commented, Irigaray's concept of porosity is predicated upon a rethinking of the relations between sight and touch (Dixon and Straughan 2010, Paterson 2004, Colls 2011). Working through the phenomenology of Merleau-Ponty, for example, Irigaray (1993) observes how the notion of an interior/exterior, self/other binary is the product, to a large degree, of a prevailing occularcentrism. When bodies are initially conceived of as sighted objects, she writes, then there is an accompanying tendency to understand touching as a means by which discrete entities are brought into physical proximity. Yet, if we prioritise touch then we must think of the body as active, already being open to touching and being touched. For Fullagar, touch is thus positioned in this relationship as a 'profoundly intersubjective sense' (2001, 179). That is, '[t]he internal and external horizons of my skin interpenetrating with yours wears away their edges, their limits, their solidarity. Creating another space – outside my framework. An opening of openness' (Irigaray 1992, 59).

How, then, does this porosity emerge within the beauty salon, a site wherein the 'state of imperfection, the unfinished condition of every living being' (Irigaray 1993, 192) is overcoded with the specification of aesthetic as well as emotional well-being as predicated on the state of one's skin? In the following I go on to flesh out the relations that make up facial treatment and the permeability that can be traced through beauty salons. Before doing so, however, I want to provide some background on the evolution of this site as a means of laying out the expectations and practices that have become associated with it.

A Social History of the Beauty Salon

Since antiquity, cultivated knowledge of ingredients has been passed down the generations and, later, across culture divides, in the creation of 'preparations' that assist in the day-to-day grooming of the body and to enhance the skin's complexion. These 'preparations', today referred to as 'products', are central to the work of the beauty salon, where they form the basis of treatments that span from waxing to manicures, pedicures, massage and facials. Designed to alter and maintain a particular physical presentation of the body, these treatments are carried out over short appointment periods that can, for example, range from slots of half an hour to two hours. While the salon's sister site, the spa, is characterised by many treatment rooms in conjunction with sauna, steam room and plunge pool, all of which panders to the body, the beauty salon tends to focus upon the head and face and be comprised of a communal space wherein numerous treatments are carried out, or a number of private spaces for one-to-one consultations.

Key to the establishment of the salon as we know it today was the emergence of the beauty industry. Here, entrepreneurs took treatments out of the home and into the sphere of marketing and manufacture. Initially centred on individual homes and locales, the beauty business nevertheless brought into its orbit druggists, department stores, patent cosmetic companies, perfumers, mail-order houses, and women's magazines (Peiss 1998, 61). According to Black (2004, 28), the industry developed as much from the enterprise 'of one woman in her kitchen' as well as 'businesswomen who developed national networks, large production capabilities'. With the development of business also came its professionalisation as training within 'systems of preferred care' (Black 2004, 28) was encouraged. Aligned in the Victorian period with negative connotations of deceit and loose morals, twentieth century marketing strategies sought to sell beauty treatments as a dutiful preservation of natural assets, as well as shared knowledge's of women's 'troubles' in regard to the health of the skin (Jacobs-Huey 2006, Willet 2000).

The 1960s, however, saw the growth of a political critique that altered the industry, as 'the modification of women's bodies [became] a personal, a public and a political issue' (Black 2004, 38). Marketing of the industry changed to emphasise a 'scientifically endorsed method of maintaining cleanliness and grooming' (Black 2004, 39), a shift that saw cosmetics become 'a science rather than an art of beautification' (Diana and Draelos 2000, vii), and the staff overseeing their use become increasingly professionalised as beauticians and cosmetologists. With the beauty industry's growth as a professional body, training and standards have become the required norm. In both the US and the UK, for example, this has been addressed through the development of beauty schools that provide modular courses, diplomas and vocation qualifications, allowing students to obtain the necessary certificates and applied skills in order to be a practicing beautician or cosmetologist. In addition to formal training there is also emphasis placed on 'hands on' experience within the salon. The US cosmetologist, for example, must undertake licensing requirements which, depending upon the state, consist of anywhere between 1500 to 3000 hours on the shop floor, plus the successful completion of the licensing exam. The UK beautician must obtain level three training at beauty school and undertake placement within a salon in order to gain practical experience. Here, then, we find the importance of an embodied, as well as conceptual, knowledge in the application of beauty practice.

Importantly, the successful trainee must also learn how to perform emotional labour. Furman (1997) for example refers to the 'nurturing' and 'self-affirming' space provided by the beauty salon, maintained by the friendship, camaraderie and emotional support of staff and customers alike. Solomon et al (2004) reiterate this association between people and place in their ethnographic study of ten beauty salons across North Carolina. The salon has become a crucial arena, they argue, within which cosmetologists and their customers talk over a wealth of topics, including health. And yet, as Sharma and Black (2001) note, this emotional labour is not reflected in pay or status, while Kang (2003, 821) points to how the 'gendered processes of physical and emotional labour in nail salon work are seeped with race and class meanings that reinforce broader structures of inequality and ideologies of difference between women'.

The modern day beauty salon emphasises its medical, scientific status through an array of products and treatments termed 'active' that are designed to enhance both hair and skin at the submolecular level. For example, the joining of pharmaceutical and cosmetic expertise has created 'cosmeceutical' products that proffer physiological as well as psychological benefits by improving the function as well as the appearance of the skin (Kligman 2000). These include a range of antioxidants, α-hydroxy acids, anti-perspirants, sunscreens and self-tanners. Designed to straddle the border between drug and cosmetic, most of these are thereby unregulated by the state (Hutt 2000). Furthermore, the salon allows the client access to treatments facilitated by equipment such as laser and light technologies, wet (botox) and dry (bioskin jetting[1]) injectables and solariums. It is important to note here that in the contemporary salon emphasis is placed on altering the structure and function of the skin through the application of technologies. Rather than focus on methods of explicit beautification, these practices try to

---

1   A process of manipulating the tissue under a wrinkle in-order to stimulate collegan growth and so 'fill out' the folds of skin.

engage clients in the scientific maintenance of bodily condition, treatments and products that clean and groom an individual. As Paulson (2008) reveals in her study of the aging body as a site of concern and treatment, the salon offers older women in particular the opportunity to confront their corporeal anxiety through consultations on issues such as problem skin, skin colour, excess or too little hair, hair condition and nail health.

Relaxing in the Treatment Room

On a summer's day in June (15/06/2009) I took an hour's respite from work and found myself lying on the treatment table of my local beauty salon in Aberystwyth whilst a beautician, who I will refer to as 'Nicola', quietly applied Dermalogica products to my face. In the treatment room I found that the dimmed lighting encouraged my eyes to close. I began to unwind as the beautician passed her hands over my face, cleansing it as they went. The window was open ajar to let a light breeze into the room. While this meant a hum of traffic from the high street was audible, the mood music that played lightly in the background detracted my attention away from it. I felt a calmness emerge.

This is not an unusual experience in the receiving of a facial treatment, for not only is the evocation provision of a sense of relaxation a central facet of customer service, but also stress reduction is key in 'skin protection'. As salon worker 'Sarah'[2] explains, 'just de-stressing and calming and looking at the things going into your body, and what you are doing to your body, reflects so much on the surface' (24/07/2009). To facilitate relaxation, beauty salons attempt to engineer affective atmospheres for clients through a number of means. This involves manipulating the senses through the use of scented candles or aroma oils, dimmed lighting, relaxing music and even, in some salons, heated memory foam mattresses so that clients feel 'cocooned'.

While 'Nicola' applied a 'deep cleanser' to take away the dead cells on the epidermis I found my sensory focus drift away from the audible music and traffic hum. 'It contains methanol traces; can you smell it?' 'Nicola' asked me. I could. I found it to be a powerful smell that had both an 'awakening' and also a 'relocating' effect. It also started to warm the skin of my face, but not uncomfortably so. Rather, the feel of the beautician's hands was now linked to a heat, as well as the distinct smell and sound; yet, my weight still pushed against the padding of the table as I lay prone. This was, 'Sarah' explained,

> … not just about a treatment, it's actually a whole mind and body experience… so its more about an experience rather than just having a certain service done. Whether it be from the beginning where we start with a sensory journey of tasting the Aveda tea, to the smell of the Aveda candles and the essential oils that

2   All of the beauticians I talked to as part of my research have been given pseudonyms.

you smell in the air, the lighting in the rooms and any of the things that you see
that you see when you walk in. (24/07/2009)

In the relative stillness of a facial, lying peacefully on a treatment bed, these
sensory techniques create 'affective resonances [that] circulate through a variety
of tactile, visual, and audio media', to produce, following Bissell (2008, 1701), a
sedentary comfort 'where the body has the capacity to anticipate and fold through
and into the physical sensation of the engineered environment promoted'. Such
mediation has the potential to create an affective atmosphere within the treatment
room, something that, Anderson suggests, 'unsettles the distinction between affect
and emotion' for they are at once 'impersonal in that they belong to collective
situations and yet can be felt as intensely personal' (2009, 80). Taking his lead
from Deleuze and Guattari, Anderson understands that atmospheres are 'generated
by bodies' that affect one another such that an encompassing or 'envelopment'
emerges (2009). An equally important point to make here is that such spatial
configurations require *work*; that is, there is an emotional labour to be found in
their actualisation, or accomplishment. 'Sarah' explained the labouring role of the
beautician in this regard:

> You could be going through anything at home but as soon as you shut that door
> and you are with your client and you turn off, and you need to be able to um,
> really pay attention to what your client is feeling and seeing ... [pause] ... you
> can instantly, the minute you shake someone's hand and talk to them for five
> minutes you can tell so much about a person, you can tell if they are really
> stressed or if they are really tired or if they just need to relax or are, if they are a
> hyper person. But if you are focused inward and you actually worried about that,
> you know, your husband or you are having problems or whatever it is, you're not
> going to notice anything (24/07/2009).

Over the past few years social geographers have, of course, turned to a number
of critical theories and philosophies to understand this kind of response-
ability in more depth. Predominantly spurred by feminist critique, care-centred
theoretical perspectives have been positioned as a 'social ontology of connection:
foregrounding social relationships of mutuality and trust (rather than dependence)'
(Lawson 2007, 3). Within this framework, care is understood to be suffused with
both emotional and affective relations (Davidson and Milligan 2004) that assist in
the reproduction of social relations. And the salon, firmly positioned between the
private and the public, the intimate and the corporate, is a site where emotional
investment and sensitivity on the part of beauticians is crucial to 'successful'
experience of the treatment room.

   Yet, if we focus on the sense of touch as it is deployed in a facial another thread of
affective facilitation can be noted, one that is woven into the 'hands on' experience of
receiving such a treatment. Still on the treatment table in Aberystwyth (06/15/2009),
the next chemical product applied was an oil, the scent of which I cannot recall, but

I found it calming. Both smooth and warming, the oil was massaged into my skin. In so doing 'Nicola' deployed a particular action as she worked her hands up, away from my neck to my chin, cupping it again and again. Breaking with this style of movement, she then followed the line of my chin towards my ears before using swirling actions around my eyes and up towards my forehead. This was, for me, a passive form of touch, or rather a *being touched*, that situates the bodies involved within particular relations. The act of being touched is, if we follow Irigaray, to create a hierarchical framing (see Colls 2011). Indeed, the act of an inappropriate touch can instil 'anxiety, fear, disgust or transgression' (Paterson 2007, 153). However, this was not my experience within this particular facial treatment, for as the beautician continued to massage in the oil, my body, especially my calves, also seemed to feel her pushing, pressing and moving the skin on my face; they seemed to tingle. It was a pleasurable, relaxing feeling, if quite intense.

As Paterson (2007, 153) notes with regards to the therapeutic touch experienced in Reiki massage practice, this can open up 'non-verbal communicative pathways between bodies that brings them into proximity'. Such proximity is, for Wyschogrod, (1981, 28) something that creates an empathetic relation between two bodies, where 'the feeling-act through which a self grasps the affective act of another through an affective act of its own'. This approach to empathy is however problematic for Paterson, as the term 'feeling-act' suggests the performance of an isolated self. Instead, Paterson refers to a sense of 'feeling-with', a phrase that invokes 'tactile empathy in a more unfolding, processual way' (2007, 167).

Whilst Paterson's (2007) discussion of therapeutic touch emerges from the affective and emotional connotations of this sense, the experience of receiving a facial creates a more literal weaving of the physical and metaphorical. For while, within the facial described here, the experience was undoubtedly a form of therapeutic touch that instilled a sense of relaxation, it was produced through a close parsing of skin on skin that mobilised this affective capacity. And again, importantly, this is a feeling-with that is achieved, or emotionally laboured for. Indeed, 'learning to touch' in order to successfully massage a client's face is not only a key facet of training, as 'Gemma' explained to me, but something that comes from a sustained, hands-on learning via experimentation upon and with the body. Thus,

> You are taught basic moves, and [beauty product companies] teach you basic moves for things.... you learn over the years ... you sort of add things that you have had yourself that are quite nice, so the moves do actually change. (21/09/2009)

Learning how to massage is very much predicated upon a *morphe* in gestation, as Irigaray puts it; in enveloping another body, one's own learns to respond via touch and, in so doing, nurture a new set of touches. As 'Anna', another beautician, said '… you are always honing your technique, you are always trying to make yourself better each time' (22/07/2009). For a beautician, part and parcel of this *is learning what it is to be touched*. Here, the hierarchical relationship between expert and client is deliberately inverted, for only by learning how touch can also unsettle can one then recognise, via touch but also by sight, the bodily clues of such a discomfort. 'Sarah' commented:

> I think if I ever owned a salon, that would be my number one thing, to make sure that my staff got to exchange and receive things because every time someone does a treatment on you, as a therapist, you are paying attention to what they are doing, and you might feel something that feels great, or you might feel something that they do that to you feels absolutely awful and it will trigger you to say, ok, I'm never going to do that to my clients. (24/07/2009)

She explained that it is important for beauticians to experience the 'feel' of touch and the use of pressures and movement so that they can build an awareness of 'great' and 'awful' feelings. And empathy here brings into relief the porosity of tactile encounters that Irigaray's concept of permeability suggests. Indeed, the porosity of action and reaction between the beautician and client experienced through tactile contact is firmly predicated upon these bodies as vulnerable; that is, open to alteration by the other. As I go on to analyse below, such a vulnerability is, in the context of the beauty salon as business, overcoded as a corporeal anxiety.

Anxiety in the Treatment Room

The massage element of a facial, which centres on what the client feels, is undoubtedly an important aspect of treatment. However, this 'hands on' aspect is also important as a diagnostic. That is, it enables information regarding the skin's texture and condition to be read. As 'Sally' told me,

> As soon as their skin is cleansed, we tone, we blot, a quick, just over with your hands ... just over the bare skin, nothing, no cleanser, nothing, so you can feel – that feels a little bit rough, that feels a bit dry, that feels a bit oily. You know, so you can feel with your bare hands what their problem is. (13/08/09)

Such non-visual modalities are usually aligned with those who are visually impaired whose spatial experience has received attention by scholars (e.g., Macpherson 2009 considers the experience of outdoor activities by blind walking enthusiasts), whilst others have considered philosophical and spatial dimensions of experiencing the world without sight (Paterson 2007). To make clear, however, the 'tactile imaginary' Sally collects is built onto both visual appraisal of the skin and through conversation. For, prior to any facial, a consultation process ensues during which beauticians take the opportunity to visually note the condition of their client's skin. Meanwhile, clients are encouraged to discuss with the beautician their concerns relating to their skin. Here in this initial stage attention is given to the visible surface in order to unpack biological structure so that a 'correct' diagnosis and helpful prescription may be applied. In addition, consultation forms are commonly filled in before any treatment occurs and these are often in-depth, designed to question lifestyle, diet, exercise and cleansing regimes so that the beautician can discern those factors which might impact upon the skin. The embodied experience of touch is thus mediated through language; in discussion, client and beautician outline the potential for the skin to feel dry or oily. Here, previously unconsidered materialities are proffered which can then be confirmed through the actual tactile appraisal.

Here, beauticians employ a mobile touch to apprehend texture; as Merleau-Ponty (2002, 367) asserts, 'the movement of one's own body is to touch what lighting is to vision'. That is, it is not in or through sight or pressure that texture can be perceived, but rather, it emerges between the grooves of skin and material. Through 'Sally's' fingers a textured pattern is relayed by dry, flaky skin fragments or their lack thereof, the raised areas of clogged pores and the slick zones of oily skin. The face is broken down into a fragmented spatiality, the fragments are rendered intelligible by virtue of their exhibition of particular skin conditions.

Yet the consultation process, coupled with these modes of touch and vision, is often not enough to unpack the materiality of a client's skin and so another technique can be brought into play, as I found out in a London salon (24/07/2009). I found that once my skin had been cleansed, the beautician placed two damp cotton buds over my eyes and moved a warm, bright light over my face. At first I questioned myself as to what she was doing. Was she *looking* at my face? My thoughts were quickly confirmed by a series of questions such as: 'do you have sensitive skin'? As I answered, a discussion ensued about my 'oily bits', blemishes and scars. Here, an enhanced, haptic visuality was brought to bear via the microscope that read my skin as a pitted, undulating surface of spotted and dimpled zones that sparked some chastisement. The gaze, so often characterised as distanciated and God-like, was simultaneously an immersion in light in this situation. Features never before suspected swarmed into view, a montage of shapes and shadows, lines and zones.

Thus problematised, the skin can be proffered special treatments such as the 'peel,' wherein 'efficacy' is more often than not understood as being manifest via the feeling of discomfort, as the following exchange illustrates,

'Sharron': I am just going to put this over the oily areas that trouble you the most. It should feel a bit tingly as well.
[there is silence for some time]
Me: Why does it feel tingly, what's it doing?
'Sharron':Where does it feel tingly?
Me: A little, just down here. [I point to my lower face area]
'Sharron': It's just exfoliating the skin, it helps to renew the skin and to diminish any um … do suffer from acne at all?

[We carry on talking as the beautician smoothes the serum around my face before using sponges to take it off].
I will just do that again ok. [She puts the serum back on] Do you find it relaxing or do you find it quite stingy?
Me: Um… it's quite stingy actually.
(16/09/2009)

As the beautician intimates, this peel is one of the salon's 'active products', suggestive in its 'feel'. Breaking down the dead skin, the chemical peel draws attention to an otherwise unregarded, if elemental, process wherein, as Goethe noted, whether 'the bark of trees, the skin of insects, the hair and feather of animals, even the epidermis of man [and woman], [these] are coverings forever being shed, cast off, given over to non-life' (cited in Benthien 2002, 85). Here we have not so much a casting away as a peeling away of the epidermis with a chemical solution that simultaneously smoothes the skin and stimulates the dermis beneath to grow. Peeling skin is, for Connor (2004, 29), associated with it's de-classification, for 'the skin is no longer a skin once it is detached. By being peeled away from the body, it has ceased to be itself'. Whereas the previous modes of skin scrutiny, both tactile and visual, fragmented the face into zones associated with particular skin conditions, here a physical fragmentation of the skin is facilitated, problematising its singularity as a 'whole' organ coating the body.

This wearing away of the skin is writ small between each procedure. And so, while still lying on the dimly lit treatment table of the salon in Aberystwyth (15/06/2009), with the smell of the menthol scented cleanser, the act of massaging was interrupted by the more grainy texture of the two sponges that were applied to and moved around my face and eyes. A sponge for each side. This texture, and its motility, was to be repeated between each subsequent treatment, as each product was applied and removed in turn: a texture between texture. Damp warm towels, cotton pads, and facial paintbrushes as well as sponges are all regularly used tactilly to wipe on and wipe off products, taking with them each time a few more cells.

An interesting way to frame this is via what Hayles (1997) terms 'corporeal anxiety,' a feeling of concern over the fate of various bodily elements, but also, it must be added, a desire to augment/enhance the same through various means. Scientific treatments that can produce a sense of self-fragmentation are constituted in large part upon the search for bodily problems, remaking identities and attenuating corporeal anxiety in the process. In the salon they are focused, paradoxically, on the creation of possibilities for smoother, clearer skin, and a youthful visage. This is predicated upon anxiety, as clients develop a deeper awareness of their corporeal fallibility. And this fallibility is highlighted by the beauticians not only in the course of questioning and scrutiny of the face, but also in the drive to educate the client of the skin's gestation. As 'Mary' told me,

I suppose one of the most important things is making sure that your client is doing what they can for their skin at home, which is where their products come

in. And people like Dermalogica and Gino sample, so you can give your clients a
sample to take home and try so they can see if that product works and does what
they want it to do. That's really important, that they don't just come in for the
facial and then go off and wreck their skin doing whatever else they are doing or
not doing as the case may be. (27/07/2009)

'Mary' indicates here how important it is for clients to continue 'caring for'
their own skin at home. The facial, with its consultation, directed products and
massage techniques offers the opportunity for the materiality of the skin to be
unpacked, such that knowledge of, and for, the client can allow for the continued
problematisation of skin. As 'Katherine' tellingly remarked, 'sometimes they
don't realise that they ... [pause] ... that [their skin] does bother them until you
start asking them in different ways' (01/09/2009). As I discovered at the end of
the facial received in London (24/07/2009), the most informative part of a salon
visit for a client can be its conclusion. On exiting the facial treatment room I was
asked to take a seat in the waiting area whilst the beautician left me to return
back the way we had come. A minute later she re-emerged with a selection of
Dermalogica sample sachets for me to take home. As she did, I noted that on the
walls of salon waiting area Dermalogica leaflets were also available, outlining
a number of ailments such as uneven skin, sensitised skin, and aging skin. The
beautician sat down next to me. Pulling out each sachet in turn from the pile she
clutched in her hand, the beautician explained what I should use them for, when
I should use them and whether or not she had used them in the facial. There were
eight sachets. It quickly became apparent that there was much more wrong with
my skin than I had realised.

Practices that narrow down inquiry to particular sites and spatialities of the
body 'reveal' to a client previously unthought of materialities in the form of
hidden problems, and possibilities to be achieved. It is a place for the multiplicities
of observation – especially the tactile and the haptic – and the development of
awareness so that materialities can be optimally transformed and maintained in
the salon and, hopefully, reiterated in the home. It is in this regard that a corporeal
anxiety can be said to animate these treatments as these bodily fragments are
scrutinised, pathologised, prescribed, and codified according to the active
ingredients that will ameliorate (but never fully erase) this problem.

Conclusion

In the treatment room, the experience of receiving a facial emerges as a cacophony
of intersubjective relations that involve the bodies of client and beautician, the
textures and feel of objects as well as the use of aroma and music. To be sure,
there is relaxation to be found here, to the extent that this is an affective situation
strenuously worked for. Further, there has been acknowledgement of the salon's
own affective capacities, and an awareness of its historical development within

the beauty industry, which has led it deeper into the scientific realm and frames it as a site for the realisation of corporeal fragmentation and consequently corporeal anxiety. In drawing on Irigaray's (1993) concept of porosity I have attempted to illustrate how the salon works with the vulnerability of the skin in the deployment of touch and the feel of being touched, such that it becomes a site where the body's 'continual gestation' is manipulated and exploited. In so doing the salon, and the skin care products deployed, dismantles the perception of a stable, ordered identity, highlighting instead something more like a corporeal fluidity and fragmentation. As this fragmentation occurs, so does the scale and intensity of corporeal scrutiny and the creation of an anxious desire to stabilise the problems manifest in each fragment. And so, while the beauty salon proffers the possibility of skin problem amelioration, it simultaneously opens up awareness of the skin's limitations and fallibility and with it, its capacity for ever further alterations, morphés, gestations.

References

Anderson, B. 2009. Affective atmospheres. *Emotion, Space and Society,* 2, 77–81.

Benthien, C. 2002. *Skin: On the Cultural Border Between Self and World.* New York: Columbia University Press.

Black, P. 2004. *The Beauty Industry: Gender, Culture, Pleasure.* London: Routledge.

Bissell, D. 2008. Comfortable bodies: sedentary affects. *Environment and Planning A,* 40(7), 1697–1712.

Colls, R. 2011. Bodies touching bodies: Jenny Saville's over-life sized paintings and the 'morpho-logics' of fat, female bodies. *Gender, Place & Culture.* DOI: 10.1080/0966369X.2011.573143.

Connor, S. 2004. *The Book of Skin.* London: Reaktion.

Davidson, J. and Milligan, C. 2004. Embodying emotion, sensing space: introducing emotional geographies. *Social and Cultural Geography,* 5, 513–32.

Diana, Z. and Draelos, M.D. 2000. *Atlas of Cosmetic Dermatology.* New York: Churchill Livingston.

Dixon, D. and Straughan, E.R. 2010. Geographies of touch/touched by geography. *Geography Compass,* 4, 449–59.

Fullagar, S. 2001. Encountering otherness: Embodied affect in Alphonso Lingis' travel writing. *Tourist Studies,* 1, 171–83.

Furman, F. 1997. *Facing the Mirror: Older Women and Beauty Shop Culture.* London: Routledge.

Hayles, K. 1997. Corporeal anxiety in *Dictionary of the Khazars*: What books talk about in the late age of print when they talk about losing their bodies. *Modern Fiction Studies,* 43(3), 800–20.

Hutt, P.B. 2000. The legal distinction in the United States between a cosmetic and a drug, in *Cosmeceuticals: Drugs v. Cosmetics*, edited by P. Elsner and M.I. Maibach. New York: Marcel Dekker.

Irigaray, L. 1992. *Elemental Passions*. New York: Routledge.

Irigaray, L. 1993. *An Ethics of Sexual Difference*. Translated by C. Burke and G.C. Gill. Ithaca, NY: Cornell University Press.

Jacobs-Huey, L. 2006. Learning through the breach. *Ethnography*, 8(2), 171–203.

Kang, M. 2003. The managed hand: The commercialization of bodies and emotions in Korean immigrant-owned nail salons. *Gender & Society*, 17(6), 820–83.

Kearns, R. 1993. Place and health: toward a reformed medical geography. *Place and Health*, 45, 139–47.

Kligman, A. 2000. Promises and problems. *Dermatologic Clinics*, 18(4), 699–709.

Lawson, V. 2007. Geographies of care and responsibility. *Annals of Association of American Geographers*, 97(1), 1–11.

Macpherson, H. 2009. Articulating blind touch: thinking through the feet. *The Senses and Society*, 4(2), 179–92.

Merleau-Ponty, M. 2002[1962]. *Phenomenology of Perception*. London: Routledge & Kegan Paul.

Paterson, M. 2004. Caresses, excesses, intimacies and estrangements. *Angelaki Journal of Theoretical Humanities*, 9(1), 165–77.

Paterson, M. 2007. *The Senses of Touch: Haptics, Affects and Technologies*. Oxford, Berg.

Paulson, S. 2008. 'Beauty is more than skin deep': An ethnographic study of beauty therapists and older women. *Journal of Aging Studies*, 22(3). 256–65.

Peiss, K. 1998. *Hope in a Jar: American Beauty Culture*. New York: Metropolitan Books.

Peterson, A. and Lupton, D. 1996. *The New Public Health: Health and Self in the Age of Risk*. London: Sage.

Rose, G. 2003. A body of questions, in *Using Social Theory: Thinking Through Research*, edited by M. Pryke, G. Rose and S. Whatmore. London: Sage.

Sharma, U. and Black, P. 2001. Look good, feel better: beauty therapy as emotional labour. *Sociology*, 35, 913–31.

Solomon, F. Linnan, L. Wasilewski, Y. Lee, A.M. Katz, M.L. and Yang, J. 2004. Observational study in ten beauty salons: results informing development of the North Carolina Beauty and Health Project. *Health Education & Behaviour*, 31, 790–807.

Straughan, E.R. 2010. The salon as clinic: problematising, treating, and caring for skin. *Social and Cultural Geography*, 11(7), 647–61.

Willet, J.A. 2000. *Permanent Waves: The Making of the American Beauty Shop*. New York: NYU Press.

Wyschogrod, E. 1981. Empathy and sympathy as tactile encounter. *Journal of Medicine and Philosophy*, 6, 25–43.

Ziarek, P. 1998. Toward a radical female imaginary: Temporality and embodiment in Irigaray's ethics. *Diacritics*, 28(1), 60–75.

Chapter 5

Fieldwork: How to get in(to) Touch. Towards a Haptic Regime of Knowledge in Geography

Anne Volvey

Introduction

This chapter touches on a central issue in geography by focusing on its relevance in the field of geographical methodology. In line with Crang's (2003) and Paterson's (2009) formative analysis, I address and elaborate the idea of a haptic regime of knowledge in geography. Continuing some of my earlier epistemological studies of fieldwork in geography (Volvey 2000, 2004) and contemporary art (Volvey 2003, 2010, Volvey and Houssay-Holzschuch 2007), and echoing some new ones (Volvey forthcoming, Volvey et al. 2012), I sustain this epistemological view of an alternative regime of knowledge with the psychoanalytical perspective of transitionality (Anzieu 1995, Roussillon 1995, Tisseron 1995, Winnicott 2005), a theory that combines self-identity and the experience of place/space through touch.

Touch is an important aspect in researchers' accounts of both traditional and new forms of qualitative fieldwork. So far, touch has largely fallen outside of the scope of the feminist, scopic-centred, critiques of conventional masculinist fieldwork practice and has only recently become a topic of more reflexive approaches to qualitative fieldwork, in which embeddedness, empathy, and embodiment became primary features from the 1990s. As to the sense of self, which arises from the witness or betweenness that characterises field geographers' situation, the issue keeps surfacing in field-geographers' accounts, but has received little reflexive attention since it cannot be reduced to positionality (social identity). Should we continue to deny the fact that researchers re-establish their self-identity by engaging with the field spatial practices? Should we neglect the roles of strong and significant haptic experiences in the building of knowledge in geography? Should we dismiss the sensations generated by such experiences as invalid data and pretend that they are not represented in geographical knowledge (construct objects and topics)? I argue here that the transitional theory can help us understand the way sense of self and experience of place/space through touch are combined in fieldwork practices as well as how these experiences are converted at a psychical level and represented.

With this transitional perspective, I participate in supplementing Bondi's (1999, 2003b) views on the potential role of psychoanalytical techniques in qualitative field-based methodologies, as well as Crang (2003) and Paterson views (2009) on haptic knowledges. However, I seek to provide a theoretical framework for the intelligibility of a haptic regime of knowledge. The achievement of this purpose requires the following: in terms of empirical geography, we must understand a regime of knowledge from the examination of the fieldwork situation (methodological procedures, practices, and experiences), we must make the researchers' fieldwork non-scientific and non-political strategies worthy of epistemological interest, we must impose a strict principle of symmetry (Latour and Woolgar 1996) in the examination of geographers engaging with fieldwork, and we must acknowledge a humanistic dimension to self-identity (Balibar et al. 2004). In so doing, we will necessarily apply the same principles of analysis we use on our researched subjects to the fieldworkers' motives, practices, experiences, field-based generated data, and knowledge constructs.

This chapter combines several perspectives, firstly, I bring together geography and psychoanalysis. By adopting an interdisciplinary approach, I engage transitionality with a geographical concern so as to combine haptic geography and psychoanalysis into a haptic regime of knowledge. In the process, I seek to shed light on the spatial turn and the haptic turn of transitionality in the field of psychoanalysis. Moreover, in psychoanalysis, I extend the relatively well-known Winnicotian transitionality of the 1960s (Winnicott 2005) to the second phase of transitional developments – French transitionality developed in the 1990s (Anzieu 1995, Kaës et al 1997). Secondly, this chapter brings together different national academic cultures: works in psychoanalysis in French and English as well as works in English-speaking geography about fieldwork and interviews with French field-geographers.

In the first part of the discussion, I concentrate on haptic issues in traditional and newer qualitative field-based geography. I discuss the researcher's questioning of sense of self that emerges from haptic and spatial experiences associated with fieldwork-as-withness. I call for an acknowledgment of their scientific relevance and for their strict subjection to the scientific principle of symmetry. Consequently, in bringing together a care-giving-based therapy and a care-giving based fieldwork, I suggest that we expand the theoretical potential of psychoanalytical transitionality – which one, I argue, underlies the methodological shift in field-based qualitative research. In the next part of the discussion, I thoroughly examine the *spatial turn* and *haptic turn* found in both the theory of the self and the therapeutic setting in transitional psychoanalysis. By developing French transitionality, which complements Winnicott's approach of constructing self-identity, and by linking the spatial features of the psychogenetic and the therapeutic processes with touch, I provide a theoretical framework that allows for the consideration of a haptic regime of knowledge. In the final part of the chapter, I address various aspects, including methodological, theoretical, and epistemological issues, of field-based geography associated with this transitional framework, providing an understanding of geography as transitional, i.e. based on a haptic regime of knowledge.

Addressing Field Touch in Geography

Touch – a concealed residue of the traditional fieldwork

To provide an account of someone's itinerary is to state one's identity. Especially where this concerns a fieldwork discipline. At first, the field appeals to the senses: it is a landscape as well as a spectacle. The piece of land on earth with its places and humans one is striving to comprehend is first captured in a sensual mode ... But the field appeals to a sensuousness that extends beyond the gaze. 'Intrusive' sounds, smells, flavours, and even tactile qualities form subtle connections that help delineate what sight has drawn attention to. Describing this sensory experience is fairly difficult ... yet there is little doubt that it influences the research process, albeit in a faint, secretive way. ... The fusion is not the sought purpose ... but an active sympathy that is an acknowledgment of the other's true humanity. In its primary and immediate expression, such sympathy doesn't actually arise from verbal exchange ... but rather from one body resonating with another through imitation. ... Research founded on categories that greatly minimise subjectivity hinders the rendering of that-which-has-been-lived. ... Sensations, emotions, the body's nostalgic memory are not discarded, however: after rubbing skin with other fellow creatures, one doesn't operate like an entomologist ... as the singular expressions of life which arouse our desire to know, belong to humankind. ... Geographical knowledge is fundamentally the concern of "*ça-voir*"[1]. What does "*voir ça*" mean? What is the "*ça*" [we] assign geographicity to? (Pourtier 1991, 92–95) [translation mine].

In this ego-geographical passage, Roland Pourtier, a French Africanist geographer and past president of the Association of the French Geographers, tackles the question of the incommensurable dimensions of fieldwork and seeks to articulate the practice's personal added value. His focus on self-identity and fieldwork – addressing a psychoanalytical matter – is significant. From an initial interpretation, this passage, which insists on the sense of sight and the suggestions of sexuality, which tends to universalise the subject's position while regarding the field in terms of possession and landscape (Rose 1993, Sparke 1996), and which resorts to Freudo-Lacanism (Rose 1993, Nast and Kobayashi 1996), seems to resonate with the feminist critiques of fieldwork as masculinist. However, is Pourtier merely relating an 'inappropriate performance of colonising power relations' (Sharp 2005, 306), modelling his work in the field on exploration, and seeking to consolidate his masculine sexual identity by surveying, penetrating with a distanced gaze, mastering, and abandoning the feminised field/space at the

---

1 A literal translation would be 'the id that sees'. This French wordplay seeks to associate an unconscious dimension (*ça* [id]) of knowledge (*savoir*) with fieldwork's sensorial dimension (*voir ça* [to see that]) so as to present a psychoanalytic perspective on fieldwork. '*Ca-voir*' is in French phonetically equivalent to '*savoir*'.

doors of the academic world (Sparke 1996)? Drawing on a passage by a white male French researcher who did research and taught in Gabon – a key location in the so-called *Françafrique* – in the post-decolonisation decades, and who later became the cross-cultural management consultant for Elf-Gabon corporation, is not mere provocation. On the contrary, I argue that there are unacknowledged dimensions in traditional field-based geography that have to date remained outside the scope of the feminist critiques, and that are worthy of examination.

Pourtier's account, which deals with fieldwork's multiple sensuous dimensions and assumes its embodiment, and – as a result – defines fieldwork's relational dimension in terms of infra-verbal communications and resonating bodies, and seeks to reveal the unconscious incorporation and transformation processes of non-visual experiences that occur within the course of field research, challenges an understanding of geographical methodology in terms of the '*scopic regime* of knowledge' (Rose 1993, Nast and Kobayashi 1996). The skin and the tactile sensations are key to the 'state of betweenness' (borrowing Nast's words 1994b, 57) – a state where the worlds of me and not-me are renegotiated – that underlies the research process described by Pourtier. These sensations generate non-visual and infra-verbal data, and Pourtier questions the scientific status and representation of such data. Furthermore, he considers this ensemble of phenomena to be part of the geographic method as well as a dimension of the geographer's 'geographicity'[2], a set of practices that he understands as a matter of self-identity endowed with humanistic dimensions. My argument calls for an acknowledgement and an understanding of fieldwork's essential biographical as well as psychoanalytical motives, which are wrapped in the researcher's corporeal experience associated with a work conducted 'with' rather than 'in' the field.

This double acknowledgement (the place of touch in traditional fieldwork and the psychoanalytical matter of self-identity) requires that, as geographers interested in the epistemology of the discipline, we focus on the emotional content of fieldwork's experience of touch, its motives, its contingencies, its elaboration into geographical knowledge (or scientific research objects), and on its ways of representation within scientific discourse. Furthermore, this questioning of the nexus fieldwork/touch/self-identity is in keeping with the current trend of qualitative methodologies, through which the fieldwork's distance, position, and spatial relationships (that is, the question of the spatialities of field and work) have been practically and theoretically redefined. I shall now examine the emergence of qualitative methodologies in geography, and discuss its already partially addressed components.

---

2   This concept was forged in French geography by Dardel (1952), and later developed by Berque (1996). Stemming from existential phenomenology, it referred first to the existential relationship to earth/space, and has become a generic term for all forms of spatiality. Pourtier (1991) uses it here to combine space, subjectivity, and signification.

Touch as the Core of Qualitative Fieldwork Methodology

In the past fifteen years, 'doing fieldwork' has received renewed epistemological attention, having become a key feminist route to understanding, presenting, and denouncing geography as a masculinist activity (Rose 1993, 1996, Sparke 1996) and having become a basis for a 'feminist political project within the discipline' (Sharp 2005, 304). The results of this critical examination have extended beyond the scope of feminist geography (see Sparke 1996, Crang 2002), providing, among others, the outlines for the revival of qualitative research methods with a wide variety of philosophical and epistemological underpinnings (Limb and Dwyer 2001). Thus, by choosing here and in the following section to articulate part of the poststructuralist feminist corpus about fieldwork from the 1990s, I do not intend to confine qualitative methodological developments to the feminist critics and new elaborations. However, by grounding their theoretical efforts in women's experiences and ways of doing, the feminists have brought into qualitative fieldwork-based methodologies notions that have placed touch at the core of the method and have designed a lexical universe worth examining. This universe sustains the relevance of epistemological consideration for a haptic regime of knowledge, and delivers interesting clues for its intelligibility.

The practice of method as a relational process and a project in acts has been crucial to the feminist epistemological turn associated with renewed anthropology (Rose 1997). The participation in the interpretative and the cultural turns in order to explore individuals' life-worlds, and the concern with the subjects' positionalities and interactions – which challenge the unequal power relations between the researcher and the researched – called for new methodologies to generate qualitative data. Feminist geographers first assumed that such new methodologies would allow the politics of the field and of representation to be transformed into political activism and reciprocal empowerment – with both the researcher and the researched resisting patriarchy through 'fieldworking' and representation (see Nast 1994a). By promoting research practices that are more intersubjective, collaborative, and non-exploitative, feminist poststructuralist geographers have portrayed the fieldworker as a care-giver and have explicitly referred to fieldwork as care-giving: '[t]he social connectedness of women to others carried out in everyday practices has fostered ways of knowing or epistemologies that are different from those of the men. Women have typically been nurturers and care-givers trained in the art of listening and other-empowerment' (Nast 1994b, 54–55). These geographers have developed an alternative feminist model to the problematic masculinist notion of fieldwork activity, and initially tended to distinguish between feminist and masculinist spatialities of the field and of the work (see Duncan 1996). Sharp (2005, 305) phrases the difference as follows: 'feminist approaches of empathy and understanding (rather than exploration, evaluation and analysis)'. By mentioning the potential intrusiveness of feminist inquiry methods' 'withness' as opposed to the masculinist distanced methods, Sharp underlines the shift in spatiality that the qualitative methods have pursued

– from 'in' to 'with' the field, and from distance to closeness (see also Rose 1996, 61). Bondi (2003b), referring to humanistic psychotherapy notions, addresses fieldworking as an empathic process (see also Aitken 2001b), while 'betweenness' – a notion that the authors of articles in the *Women in the Field* theme issue (Nast 1994a) have implicitly borrowed from Winnicott – has become common in addressing the politics of positionality (Rose 1996, Sparke 1996, Cupples 2002, Powell 2002). Unsurprisingly, drawing on this caregiving-based notion of fieldwork, the idea has arisen among feminists, and more generally qualitative geographers, that the research process is not merely dialogic but also embodied (Nast 1998, Parr 2001), is not only practiced but also corporally performed (Rose 1997). The body is then considered an actual 'instrument of research' (Longhurst et al 2008, Paterson 2009), i.e. an 'active agent in making knowledge' (Crang 2003, 499) as well as a 'site where the difference is placed' (Nast 1998, 95) – a key marker of the researcher's positionality. The research process depends on the body in movement and action (Paterson 2009) and takes place in the framework of a setting designed by co-present bodies that interact physically, symbolically, and emotionally (Punch 2001, Aitken 2001a, Longhurst et al 2008, Sotelo 2010, Lawrence 2010). The exploration of the life-worlds of individuals and the new considerations relating to the role of the body in the research process have helped expand fieldwork to sharing the lives of research subjects (Aitken 2001b, Lloyd et al. 2012). Punch (2001) reports on her preparing meals and sleeping in the same room with families, as well as playing with children in southern Bolivia. Longhurst et al (2008) have made this new situation the foundation of their research methodology when they describe a shared lunch with migrants in New Zealand. Hence, by 'listening to, giving voice to and representing the silenced' (Crang 2002, 648), and by considering fieldwork a discursive process, feminist geographers have long chosen verbal over visual procedures and material (Crang 2003). Nevertheless, qualitative researchers have started recognising that fieldwork's corporeal and emotional experiences (above all, the tactile experiences) are likely to provide the researcher with 'pre-verbal' data (Aitken 2001a, Bingley 2003, Bondi 2003b, Crang 2005). Touch is considered as the operator of an engaged fieldwork-as-'withness', and the processor of a performative and proximal form of embodied and empathic knowledge, which Crang (2003, 2005) and Paterson (2009) call 'haptic knowledges'. The question has become: how do geographers represent the haptically generated qualitative data (Longhurst et al 2008, Paterson 2009)?

Beyond Touch – Addressing the Sense of Self

The journey I present here through qualitative epistemology of fieldwork developed in the last decades, its trajectory from a *discursive knowledge regime* – that replaced classical epistemology's *scopic regime* – to a *haptic regime*, again encounters touch. Touch emerges from the forms of fieldwork engagement that feminists have forged in theory and in practice, and from the subsequent shift in spatiality they have achieved within the framework of qualitative research. In fact, this development in the epistemology of fieldwork is turning touch – a peripheral issue in classical fieldwork – into a central aspect of the reflection on fieldwork for qualitative researchers. According to Crang (2003, 2005) and Paterson (2009), along with sight and discourse, touch is a 'form of knowledge' in making place/space. I argue that the increased centrality of touch is associated with the reinforcement of the researchers' questioning of their *self*. I shall now address this matter, which has not yet been articulated clearly, and define the self-identity it refers to.

> Fieldwork for Western-trained academics is foremost a private, inner exploration. … Selfishness is inherent in individuated Western fieldwork in non-Western settings, an arrogant assumption that somehow one person develops explanatory powers. One quickly learns that selfishness must give way to a sharing, an open-ended identity enmeshed in a community. This giving way of self, however, also has its limits. … To negotiate the boundaries between selfishness and selflessness addresses how I go about 'casting within [my]self the scalpel that carries out [my] separations' as Kristeva put it. … In every research experience though, it remains a constantly negotiated boundary space within me and around me, as well it should (Myers 2001, 193, 196).

Self-identity is a concern for field geographers doing qualitative research, as Myers (2001) alludes to when moving from the political issue of 'selfishness' to the question of the sense of self, when evoking the negotiated boundary between the me and the not-me as well as when, in another part of his paper, he ascribes psychoanalytical motives to Western fieldworkers' 'search for the self'. Furthermore, he suggests a borderline situation about his self in his play on his two first names Garth and Andrew, in his focus on the environment's physical features and events in order to depict the limits inherent to his 'selfless embeddedness' and the grounds of his renegotiated separated self. References to limits are central to Myers's accounts of his borderline self-experience: either consolidated – locks, 'burglar bars' on windows and doors, and 'shards of glass' on top of courtyard walls – or, at the contrary, deficient – his house as the site of a 'constant stream of houseguests and visitors', his bedroom walls being the conductor of constant and invasive conversations – spatial limits function as a metaphor to depict the successive hardening and blurring of the boundary of his self as well as resources (both physical and symbolic) to support the negotiation of his sense of identity. Myers's sense of self relates to his experience of place/space, and its negotiation depends on evolving spatial relationships and evolving uses of the spatial situation.

Sharp (2005, 307) also addresses field researchers' needs relating to their sense of self, and narcissistic suffering, when, having discussed the 'embodied challenges of the field (often physically or emotionally overwhelming at the time)', she echoes the calls for a 'need to ensure there is space for the 'care of the self' in addition to paying close attention to the ways in which research subjects are treated'.

Fieldwork … it begins when you tear yourself away. … Actually, the field is a playground. … It is a place where you take risks. The risk is to let your personality go, to lose yourself, to lose your life. … In fieldwork, your body is at stake, your psyche is at stake … and it can be dangerous. Yet it is truly '*jouissif*' [exhilarating]. … It is the same sensation as surfing. You are standing on the surfboard, you know where your marks are, and suddenly the wave takes you … you get the fright of your life, you are lost. And then, suddenly, within a fraction of second, you breathe, and then you see the sky, you know where the bottom is, where the top is … Somewhere, integrating the air, you become sure of your marks. … What you also experience [in the field] is being separated. Somewhere, you build your self, you build the other, and – in so doing – you build this something between the two. [translation mine]

In 1991, when I came back here [in the field], a friend told me 'come to my place, I have an extra room'… In fact she gave me her room, but … shortly afterwards, she came back to her room, where she had a double bed, and where discussions between us took place. Finally, … she resumed sleeping in her bed, with me. Since then we have always slept in the same bed, even when working together in another village. … It is true what I've already told you: when I returned [to France], I missed it. When I go back there, I like this sleeping-together situation. I don't know, it is reassuring, it has a dimension… this strong closeness that has been established, this integration, is hard to express. It is a feeling that is not at all sensual, but emotional. It is affectionate … it is the feeling that you are inside. … I leave clothes there, since I have two parkas, two snowpants … I maintain a presence there. And when I am away, I tell her 'you can wear my clothes if you want to', so she wears my parka. When I come back there, it is a real pleasure to put on these clothes, to be in them. When you wear these clothes, you are not exactly the same as when you wear other clothes. [translation mine]

These two lengthy quotations are a forty-year-old male and female French ethno-geographers[3]. I believe they are significant because they evoke both Pourtier (1991) and Myers' (2001) sentiments and they indicate a continuity of concerns in geographers' accounts of fieldwork (despite age, gender or academic culture differences). The sensuous experiences they narrate depend on spatial relationships and doings, and are processed through touch or empathy; they pertain to playing, they trigger overwhelming alternate feelings of fright, comfort, pleasure, *jouissance*; they provide access to an existential sense of self that is not easy to represent.

This questioning of identity in terms of 'search' and 'care of the self' cannot simply be reduced to the researcher's positionality – even when reflexivity is about the political locatedness of the body (Nast 1998, Longhurst et al 2008).

---

3   They are abstracts from thirty in-depth interviews with French geographers (some qualitative researchers, others not) I conducted in 1999 and 2000. In these interviews, I sought to study researchers' fieldwork practices and experiences (Volvey 2000, 2004).

The positional understanding of reflexivity based on a spectrum of categorical differences (class, race, gender, age, disability, etc.) is not likely to deliver such an existential motive. Neither knowledge nor empowerment are at stake here. Fieldwork-as-withness is a borderline situation in which researchers' narcissistic dimensions (both suffering and re-negotiated) are at stake – a borderline situation acknowledged but often resisted[4]. Thus we need a humanistic definition of self-identity (see also Bondi 1999) – as opposed to a social definition of it –, a narcissistic definition that articulates subjectivity with egoity, and renders the subject's egoic dimension (Balibar et al. 2004). Should we deem the sensations generated by such experiences as invalid data and pretend that they are not represented in geographical production? The expanded fieldwork I referred to earlier (Aitken 2001b, Punch 2001) argues for an enlarged understanding of cognitive process and embodied data that would encompass the researchers' sensuous and emotional experiences of being with or sharing lives.

The 'search for the self', on the one hand, and the suggestion of 'care of the self' as well as the idea that there could be a fieldwork-spatial strategy of self-caring, on the other hand, should be taken seriously and examined epistemologically. The elaboration of these notions calls for a theoretical framework that is commensurable with the corporeal and emotional experiences attached to qualitative methodology practice and with its psychoanalytical perspective.

Issues in Haptic Geography

The key arguments in the passage by Pourtier quoted above – touch (haptic experience), non-visual data, self-identity, and withness – challenge the Freudian-Lacanian-based feminist criticism of the unconscious motives and experiencing conditions of masculinist work in the field. This criticism articulates the politics of the field with visual practice and interprets it as a matter of sexual identity. On the other hand, embodiment, touch, and non-verbal spatial knowledge have become keys to qualitative methodologies, while the fieldworker's self-identity engaging with the field is an emerging topic. I therefore argue that there is something about the fieldwork-as-withness/touch/self-identity nexus that must be considered, epistemologically elaborated, and articulated in geography. This involves, first, the strict application of the principle of symmetry to working

---

4 Myers (2001), for example, not only opens his paper with a quotation by Kristeva about the strategy of place and privacy, he also refers to the 'various modes of psychoanalysis' contributed by some geographers. However, denying a desire to 'sidestep' such an 'opportunity', he chooses storytelling about himself over psychoanalyzing his field situation. Myers's acknowledgement of a fieldwork psychoanalytic motive and his resistance to it echo Pourtier's comments at the 'Journée de l'AGF' in December 2006: 'Fieldwork is *'ça voir'*: the *'ça'* [*id*] I find somewhere on the field, but let's return to more scientific topics' [translation mine].

'with' the field, where fieldworkers' embodied experiences of touch and the generated haptic knowledges – both attached to their practice – are considered valid subjects for epistemological examination. In my view, this is not mere reactionary solipsism, as researchers' doing-with-space/place and geographical knowledge are subjected to the same principles of analysis we use on our research subjects. This also means that we would utilise theoretical tools that would help elaborate such a nexus as well as the ways geographers' haptic experiences are processed into geographical representations.

Crang (2003) has addressed the matter of haptic knowledge in qualitative methodologies, while Paterson (2009) – answering Crang's call for theory – has surveyed the matter's relevance in contemporary social science and geography literature, has provided the post-phenomenological and non-representational theoretical background of touch and 'sensuous dispositions', and has emphasised the tricky matter of representation. In so doing, Paterson has contributed to consolidating sensuous and tactile 'geographies' as a topic of geographical science. Nevertheless, neither Paterson nor Crang mention Bingley's (2003) design of her fieldwork technique according to Winnicott's notions of 'facilitating holding environment', 'potential space', 'transitional object' and 'play' in order to access the tactile experience of sense of place/landscape. Thus, Winnicottian references remain outside the scope of their survey of 'feely-touchy methods' and their theoretical (post)phenomenological and non-representational foundation of their notion of haptic knowledge. Furthermore – despite his concerns for 'sensuous scholarship', for embodied fieldwork procedures[5] – Paterson (2009) has limited the scope of their relevance to researches that deal explicitly with haptic situations (e.g., life-worlds of the blind and visually impaired, experiences of nudity on the beach, of a body that suffers or feels pleasure) and has not elaborated a haptic regime of geographical knowledge that would encompass the researchers' experiences. If, as revealed by Paterson's (2009) account of specific experiences, spatiality also derives from the experiences of touch, and if, according to Crang (2003, 499), an examination of 'the actual processes of learning through our bodies' responses and situations' in qualitative research consolidates the scientific relevance of 'touchy-feely methods' and haptic knowledges, we must also articulate the haptic experiences of field-geographers in relation to their practice so as to address the constitution of geography (as a scientific knowledge) through touch.

Hence, I assume that feminist geographers have opened the way to the examination of the mentioned nexus by resorting to two different sets of psychoanalytical theories to criticise or, conversely, sustain empirical methodologies' motives, experiences, and geographies. As noted, feminists have explicitly resorted to Freudo-Lacanism to politically examine the traditional masculinist fieldwork as well as to Winnicott's notions of the therapeutic setting to theoretically and practically develop their

---

5   For example, 'performing ethnographies of the somatic sensations within walking' (Paterson 2009, 777).

research stance in terms of 'empathic process' and 'withness' or 'betweenness'[6] (Nast 1994b, Rose 1996, Bondi 2003b, Bingley 2003). Bondi (1999, 17) notes that 'human geography has forgotten or neglected links with the humanistic psychotherapy movement ... especially in texts concerned with methodological practice', before finally turning to the works of Rogers (1957) and Winnicott 'as resource[s] for reflecting on fieldwork relationships negotiated in the course of feminist geography' (Bondi 2003b, 64) or for reflecting on the emotional dimensions of fieldwork encounters (Bondi 2005). More generally, Winnicott's transitional notions have served as conceptual resources to back up poststructuralist feminist reflections on qualitative methodology, providing the grounds for a transitional framework for fieldwork qualitative practice (cf. Aitken and Herman 1997, Kneale 2001, Bingley 2003). Thus, a care-giving based therapy has become, explicitly or not, the operational model for a care-giving based methodology in geography. However, if feminists have resorted to Freudo-Lacanian psychoanalytical notions to denounce a sexual identity that – in their view – constitutes the unconscious motive of masculinist politics of fieldwork and geography, they have not – to my knowledge – examined the epistemological implications for fieldwork and geography of the core issue of the psychoanalytical framework on which they have founded their methodology, the sense of self-identity. For instance, Bondi (1999, 18) brings together the patients' increasing 'needs pertaining to their sense of self' and the development of humanistic psychology, and – when addressing the role of empathy and unconscious processes of identification in qualitatively generating data – she mentions the dimension of fieldwork's personal added value. However, if Bondi (2003b, 67) acknowledges the potential therapeutic 'side-effects' of the research relationships for the 'research respondent', she fails to consider the fieldworker's self needs, motives and benefits. Seeking to establish a clear distinction between the practices of the therapist and that of the qualitative researcher, she does not apply the principle of symmetry to the consideration of fieldwork relationships and personal benefits, which – beyond methodology – would help us consider the epistemological relevance of transitional processes and experiences. While geographers have addressed 'what being there in the field means in terms of the production of knowledge and producing the authority of the researcher' (Crang 2003, 495), they have not addressed what the elaboration of a care-giving based methodology and the implementation of transitional processes in fieldwork mean in terms of producing the researcher's self-identity as well as their epistemological implications for geography.

I now propose that *transitionality* – a psychoanalytic theory that, drawing on Winnicott, has built a spatial and haptic theory of the self and of the narcissistic sufferings' clinical setting – should be employed as a tool to consider a haptic

---

6   In the quasi inaugural *Women in the Field* theme issue (Nast 1994a), the general ideas and the words themselves –'state of betweenness', 'space of betweenness', 'negotiating the worlds of me and not-me', 'playing in the field' (the title of Katz's paper) – are implicitly related to Winnicott's (2005) notions of therapy setting and process.

regime of knowledge in geography (Volvey et al 2012). This haptic regime should include the research motives, conditions of possibility, generated data, and their representations. Therefore, I do not simply consider transitionality as a theory that helps one examine qualitative methods reflexively, I aim at using it as a theory that helps one understand and elaborate an alternative to the scopic regime of knowledge. I shall first present a theoretical reading of transitionality. On the one hand, I seek to bridge English and French transitionalities so as to outline the haptic constitution of the self-identity and on the other hand, I focus on the relevance of spatiality in transitionality so as to reveal the link between sense of self and experience of place/space through touch.

Transitionality: Establishing the French Connection

Presenting transitionality

In the past fifteen years, transitional psychoanalysis has been explicitly introduced to English-speaking geography (Aitken and Herman 1997, Bondi 1999, 2003a, 2003b, Bingley 2003) as well as French-speaking geography (Volvey 2000, 2004, forthcoming). The formers associate transitional psychoanalysis with Winnicott, the founder of so-called 'transitionality' in the 1960s (due to his main concept, 'transitional space'), along with weaker ties to John Bowlby (1969) and his theory of attachment. Furthermore, the spatiality of transitionality has either been discussed mostly at a metaphorical level (for example, in Aitken and Herman 1997) or has deliberately been confined to a methodological level (Bondi 2003b, Bingley 2003), if not to counselling (Bondi 2003a). Beyond geography, transitionality is generally considered a set of therapeutic methods gathered under the general concept of 'holding' – introduced to psychoanalysis by Winnicott. However, transitionality is more than a set of procedures; it is an alternative theory to Freudo-Lacanism that is no longer limited to Winnicott's work.

In the 1990s, French scholars of transitionality refer to Winnicott in order to oppose the formalism of Lacanism (Roussillon 1999, 11, 14, Clancier 1999, 202). They turned these references into a paradigm shift issue, resorting to a history of mental disorders[7] to ground it, formulating it to sustain their own practical and theoretical matrix (Anzieu 1995, 28–29, 31, Green 1999, 172, 176, Golse 1999, 14). The French psychoanalyst Green (2000) calls this notion of building self-identity and its possible failures a 'theory of touch'. Anzieu, Tisseron and Roussillon – the most relevant scholars in this regard – have developed a wholly spatial (Anzieu 1995, 33, Roussillon 1999, 24), corporeal (Anzieu 1974, 195), cutaneous (Anzieu 1995, 119, Golse 1999, 14, Green 2000, 221), and infra-verbal (Anzieu 1956, 228–231, Green 1999, 173, Golse 1999, 14) alternative to (post)Freudian time, sex (erogenous zones), and language-oriented notions of psychoanalytic processes.

---

7   From neurosis to narcissistic-identity conditions via psychosis.

Furthermore, they have established their alternative in the theoretical field of individuation/subjectification processes as well as in the theoretical field of the therapeutic setting. This new, psychogenetic approach with phenomenological underpinning endows spatial and corporeal dimensions with self-identity. Although this is an acknowledged paradigm shift situation (Chabert 1996, 75), it has not yet been properly articulated in the field of psychoanalysis (Volvey 2004). Indeed, Winnicott (2005, 148) provides a paradoxical definition of 'transitional space' as a 'construct that negates the idea of space and separation between the baby and the mother', however he uses various formulations (including references to space) to depict the transitional process, as his French followers did too.

A spatial turn in the field of psychoanalytical theory

Winnicott considers that psychogenesis is a fluid sequence of states that depends on the ongoing use a baby makes of an evolving 'facilitating environment' that is kept continuous but also continuously altered by care-givers. These are states rather than (post)Freudian stages because the individuation/subjectification process is an experienced one that corresponds to various ways of being; because it is also non-linear – i.e. made of three different moments ('narcissistic', 'transitional', and 'cultural') that are all experienced anew under specific facilitating circumstances. The environment as a whole is an ensemble of different cares and different spatial arrangements, articulated within the three-way framework of holding, handling, and object-presenting. This setting evolves over time, along with the mother's cares, in response to the infant's needs and plays with the mother's cares[8]; it supports the infant's 'journey' from union with the care-givers to separation from them.

The postural arrangement of bodies evolves according to variations in the spatial relationships between the care-givers and the infant. Spatially, this varying posture leads the infant from the experience of a vertical arrangement of place (implying no corporeal distances between infant and care-givers) to the experience of a horizontal arrangement of place (implying corporeal distances). Infants' use of an object presented to them (e.g., parts of their body, parts of the care-givers' body or their entire body, or animate or inanimate objects) endows a spatiality, and also evolves. Although their 'created (inside)/found (outside)' paradoxical 'playing' within the frame of the setting takes place on a psychical level, it is based on acts of (dis)placement. As Winnicott (2005, 55) puts it, 'playing is doing' and 'involves the body'. The uses of objects evolve from the 'magical control' of the 'subjective object' to the 'destruction' of the object, and its recognition as 'not-me' when it happens to 'survive' its destruction.

Thus, one can understand the spatial dimension of the terms used by Winnicott (2005, 136) when he qualifies the transitional construct: *'ego-relatedness*, at the place where it can be said that *continuity* is giving place to *contiguity*'. Transitional

---

8   The cited scholars use 'mother' in an abstract sense, i.e. less in reference to the biological mother than to the primary caregivers or *entourage* (Anzieu 1995).

space is a symbolic construct that corresponds to the moment when infants' psychical (i.e. imaginary) construct of the union with their environment is distinguished from their physical (i.e. sensorimotoric) experience of separation from it as well as facilitates the building of a psychical basis for actual relationships between the self and external reality. Furthermore, while the 'transitional object' that represents the transition can move from place to place on a sensorimotoric level, it still relates to the environment on the level of imagination: the infant can psychically lean on the transitional object while physically experiencing the actual separation.

Hence, an infant draws an 'internal environment' from the psychisation of its body experiences associated with the changes in both the doings (caring/playing) and the spatial arrangements; as a result, this pattern founds the sense of self. Every evolving modality of the setting (i.e. the doings and the spatial arrangements associated with holding, handling, and object-presenting) is psychically worked out (playing) and yields a specific modality of 'internal environment'. The 'transitional state' (transitional space and transitional object) follows the 'narcissistic state' ('anaclitic position' and 'subjective object') and precedes the 'cultural state'; i.e. the three states comprise the infant's 'journey' to subjectivity and sense of identity. This journey supports the infant's feeling of being a united entity, i.e. contained within the limits of its body, psychically integrated, personalised, and endowed with the ability to have relationships with self, others, and the world outside as well as to act reflexively.

Touch in the theory of self-identity ('Au fond de soi le toucher', Anzieu 1984)

While Francophone thinkers of transitionality draw on Winnicott's ideas, they supplement them with an understanding of psychisation processes. Their reinvestment in anaclisis in terms of attachment and their definition of a non-language-oriented process of symbolisation render their proposal a *theory of touch* (Green 2000) founded on consideration of skin, rather than body. Anzieu, Roussillon, and Tisseron endow it with a special role (Golse 1999, 14); the skin is considered the instrument of infants' experiences associated with the settings as the infant's processor and converter of the experienced data, but also as the pattern of the symbolisation process. As Green (2000, 221), commenting on Anzieu, asks: 'does skin, the origin of perception, give rise to representations in the same symbolical way the anal object would? We may well believe this' [translation mine].

According to Anzieu (1995), an infant and its *entourage* are bound by exchanges processed by and mediated through the skin that occur in the care setting via body-to-body contacts. Anzieu calls this ensemble of touches that conveys messages and meanings 'echotactilism' and extends it to 'echolalias', 'echorythmia', etc. in an attempt to outline its enveloping structure (a double-centred inclusive structure built upon *en miroir* exchanges, one imitating the other). Skin – humans' largest sense organ covering the entire body – is considered a surface of perception, communication, recording, and elaboration; it processes the cutaneous data associated with touch. Infants perceive sensations on their bodily surfaces; these sensations are elaborated into 'psychical representations' [*figurations psychiques*] and are worked on at the level of imagination. Anzieu (1995) calls these representations 'skin-ego' [*moi-peau*] and places them at the basis of self-consciousness.

Beyond the metaphor, Anzieu (1995) proposes an analogy between the skin and the ego; in other words, an identity in structure and function. From the enveloping of their body that infants experience in the care setting, they draw a psychical pattern of their ego: a scheme of its organisation and functioning. Infants use these 'sensation-images' [*images-sensation*] to build early figurative self-representations as an ego with psychological content. The pattern of skin-ego psychical elaboration is the biological functions of the skin. Due to an unsupportive *entourage*, process failures are responsible for 'formal signifiers' [*signifiants formels*] (signs of identity conditions). While the building process is unconscious, the various images of skin-ego [*figurations*] are pre-conscious or conscious. According to their definition as 'intermediary structures' (between the care-givers and the infant, between fusion and separation), the various images of skin-ego belong to the set of transitional phenomena, complementing the transitional space and object as transitional constructs[9].

In order to explain the transposition at a psychical level of sensations associated with the skin's biological functioning experienced in the care setting, Anzieu resorts to Freudian concepts: anaclisis [*anlehnung, étayage*] and drive [*trieb, pulsion*]. Anzieu completely rethinks these concepts from the relational perspective of transitionality (Cupa 2000, Volvey 2004, Roussillon 2008a) and reloads their spatial and corporeal understandings. First, Anzieu works with a double anaclisis: anaclisis on the 'mother-environment' (enveloping maternal care) and anaclisis on the skin (the enveloped infant body). Anzieu also develops the notion of attachment suggested by Bowlby, turning its etho-behavioural features into a psychoanalytic notion. According to Bowlby, attachment is the behaviour of infants seeking to maintain physical proximity between themselves and one or various attachment figures, and which responds to the acts of care. According to Anzieu, attachment is a drive and therefore a borderline concept between soma and psyche. Nevertheless, attachment represents the relational perspective (Roussillon 2008a) as well as its spatial grounds (Volvey 2004). This energy finds its source of excitement in the infant's skin, which is stimulated by the care setting (arrangements of places and doings) and its first object of cathexis in the mother-environment. Infants attain their objectives when the spatial connection with the object of cathexis is achieved in the anaclitic situation, which makes echotactilism work – the energy provides the infants with an imaginary basis, the 'common-skin fantasy' [*le fantasme de peau commune*], that props up the ego's integrative and reflexive momentum. Drawing on the reflexive structure of touch, the infant's attachment drive now has

---

9   '[Skin-ego] … corresponds to the moment when the psychical ego differentiates itself from the corporal Ego on an operative level, and is still confused with it on a figurative level' (Anzieu 1995: 61) [translation mine].

a secondary momentum, auto-reflexive, and a second object of cathexis, its skin, which leads to a second type of fantasy – the 'skin-ego'.

Given the relational nature of transitionality, these processes are facilitated by the signs sent or echoed by the *entourage*. It is not only the fantasy of union with the care-givers that is supported by the echoing situation, but the drive's auto-reflexive process that depends on the 'prohibition of touch'; the *entourage* articulates and enacts when its response to the infant's needs is altered and when changes occur within the doings and the setting's spatial arrangements. The process of differentiation therefore depends on both the infant and the *entourage*, whose differentiated and rhythmic ways of offering itself to the infant favours the infant's fulfilment of experiences of proximity and distancing, fusion and separation. The *entourage* is considered a malleable resource situation whose doings and articulations of alternating proximity and distance are perceived, processed, and converted via the touched skin in the infant's early symbolisation work. The *entourage*'s many ways of managing by doing and articulating the setting's spatiality in response to the infant's manifestations of attachment (an ensemble of acts and sounds that seeks to maintain proximity) support the emergence and the evolution of structuring fantasies.

Therefore, the skin-ego is representative of the attachment drive conveyed to the psyche by the soma under certain relational, spatial, and corporeal circumstances. The shift in the drive's object from the mother-environment (enveloping maternal care) to the infant's skin (the body envelopment) initiates the creation of a 'narcissistic envelope' in the psyche and psychologically makes possible the separation between infant and *entourage*, i.e. it allows the 'common-skin fantasy' to be erased. At this moment, infants recognise at a symbolic level that they and the attachment figures have their own skins and egos, which contain thoughts, fantasies, and emotions, and are located on either side of a limit that has become a medium for communication. This rethinking of the attachment drive and of anaclisis in terms of touch restores their spatial and corporeal dimensions to psychical processes.

Space and Touch in the Theory of the Therapeutic Setting

Séchaud has emphasised that transitionality first addresses the therapeutic setting (cf. Anzieu 1995, 4). Having encountered an ensemble of new pathologies (i.e. identity disorders) since the late 1950s and early 1960s, transitional psychoanalysts have pondered the conditions of possibility of their therapy and have contributed to the theory of the therapeutic setting (Kaës et al 1997). They have gone from metapsychology (the theory of psychoanalytical experience) to psychoanalysis and have designed a relational, spatial, and corporeal theory of self-identity rooted in psychoanalytical practice. With transitionality, psychoanalytic theory has made its *spatial turn*.

This concern with *space* is not merely metaphorical because, by abandoning dream-based Freudian therapy for care giving based therapy (Roussillon 1995,

Bleger 1997)[10], transitional therapy acknowledges a place for the body in the psychoanalytical situation and understands the psychoanalytical process as spatial relationships between co-present bodies in interaction. The 'psychoanalysing space' (Roussillon 1995, 24) is an 'intermediate area of experience' (Winnicott 2005) in which spatial relationships and dimensions can be played with to resume – via the embodied experiences associated with playing – the psychogenetic process at the place where it first failed. Consequently, the transitional setting is a spatial situation where healing occurs, rather than a time of understanding. The spatial grounds of transitionality are responsible for the success of the concept of 'holding', which has become the main metaphor for this renewed therapeutic situation (Khan 1971), while – conversely – *touch* has become a problematic issue. Holding, which renders an account of both the infants' and care-givers' doings (playing/caring) and the arrangements of place, has also seen the further development of the transitional setting towards group therapy (Kaës 1997), analytical psychodrama (Anzieu 1979, Kaës 1997), art therapy (Lecourt 1987), the technique of pack[11] (Delion 1998), and so on. Distinguishing itself from the so-called emotional or humanistic psychotherapies (for instance, Delourme 1997, Prayez 1994) and seeking strength in the history of psychoanalysis, and in its understanding of the psychogenetic process itself, French transitionality reasserts the Freudian rule of the 'prohibition of touch' (Anzieu 1995, 166). Symbolic touch versus actual touch in the transitional setting of Mrs Oggi's therapy is discussed by Kaspi (1997) and Anzieu (1997, 196–98). Anzieu (1995, 1997), who circumvents the 'prohibition of seeing' by developing a face-to-face setting, complies with the 'prohibition of touch' by advocating work with enveloping sounds and words as well as through empathy[12]. Nevertheless, Anzieu (1989, 113) acknowledges the supportive effect of what he calls 'the therapy of relief envelopes' (encounter groups, massages, bioenergetics, psychoanalytical relaxation, etc.) on 'individuals, who need periodically to reconfirm through practical experience their basic sense of having a skin-ego'. While emotional or humanistic psychotherapies that claim relationships to transitionality resort to actual therapeutic touch, contradicting Anzieu's mischaracterisation of touch as sexual acting out (Prayez 1994).

From this therapeutic perspective, Serge Tisseron's (1995, 1996) contribution to transitionality refers to his explanation of a 'psychical imaging activity' [*activité psychique imageante*], which – based on drawing and modelling – informs his psychoanalytical technique. While Tisseron's explanation is similar to the notions of Winnicott and Anzieu, he seeks to combine them through 'psychical

---

10   The psychotherapist echoes early care-givers in order to enable the patient to play with the multiple dimensions of the setting.

11   The technique of pack corresponds to wrapping patients in sheets and doubling them with close encirclement by the group of medical personnel.

12   This technique of empathy is based on the French use of the verb *toucher* (to touch) on an emotional level, meaning to affect or to be moved. Anzieu argues that statements or enactments by the analyst and the patient can work as symbolic equivalents of actual touch.

imaging activity' that helps organise and convert touch experiences into primary self-representations. Tisseron calls the self-representations 'schemata' – 'of transformation' (union and separation), 'of envelopment' (containment) – and calls their concrete constructs 'schemata images' [*images de scheme*]. He argues that the drawing situation – which depends on a body position, a performance, and a sheet of paper – is an introjected setting that holds the psychisation of the individuation process: the opening of a field of activity by the act of drawing on a supportive piece of paper presents a background that encompasses the symbolisation process of ego images. Tisseron does not really consider the intention of representing an object that drives the act of drawing, but rather the complex of sensations, emotions, and motoric functions involved in the act of tracing when held by the setting. He examines the way in which this experienced situation facilitates the healing of failed self images. 'Schemata images' rise up from and are outlined against the background: the traces on the paper are signs of the drawing situation's use as a process of psychisation of touch and spatial experiences. Within therapy, this technique resembles the squiggle game[13] developed by Winnicott with young children in order to institute therapeutic holding.

In the third section, I address the relevance of transitionality in analysing fieldwork-as-withness processes in terms of touch, space, and sense of self, so as to outline the grounds of a haptic scientific regime in geography.

Addressing a Haptic Regime of Knowledge: Geography as a Transitional Science?

Reading transitionality in the context of fieldwork-based geography

I shall first outline the elements that support the relevance of transitionality for founding a haptic regime in geography. Thanks to Winnicott's relational and phenomenological perspective, we recognise the role of the care-givers and of the infant as well as the key role of *space* (the setting's spatial arrangements) and *spatiality* (the actors' performances with space) in the individuation/subjectification process's conditions of possibility. Transitional space is a symbolic construct that organises a certain state of spatial relationships that is physically experienced by infants and prepares the emergence of their sense of self. Spatial relationships (with the setting, between its actors) function as resources for the process of individuation/subjectification, and can be used/played again within the therapeutic framework. The *sense of self* is bound to the *experience of space/place*.

In turn, Anzieu's (1995) notions of double anaclisis and an attachment drive allow us to recognise the role of the experience of touch in the construction of

---

13 'In the squiggle game I make some kind of an impulsive line-drawing and invite the child whom I am interviewing to turn it into something, and then he makes a squiggle for me to turn into something in my turn' (Winnicott 2005: 22).

primary self-images. Anzieu also provides a description of the psychical activity in terms of 'psychisation' which stands in opposition to 'mentalisation' as to render the role of the corporeal and figurative processes as well as to oppose the idea of an always *déjà-là* autonomous psyche (Golse 1999). The primary symbolisation process described by Anzieu is infra-verbal, figurative rather than logo-dynamic, and its constructs are body images. The symbolisation process of touch-generated data he describes is less a somatopy than a somatopography, which gives way to a *somacartography*: the cutaneous data are surveyed in such a way that they inform subjects' narcissistic dimensions – in Winnicott's words, their 'internal environment'[14]. This calls us to update Anzieu's skin-ego lexicon in terms of mapping (as opposed to imaging), i.e. 'mapping of sack', 'of screen', 'of sieve'. With Tisseron's therapeutic technique, we observe the importance of the drawing or modelling situation that enables the psychisation of haptic experience (i.e. mapping) to resume under facilitating circumstances. Tisseron's non-representational view of drawing and modelling (as both process and tracing) is significant because he articulates a spatial, corporeal and non-verbal technique that helps generate haptic data and convert them psychically, beyond Anzieu's view of skin-ego mapping. Anzieu's as well as Tisseron's primary or secondary transitional constructs are endowed with comparable spatial features (Anzieu's surface, limit, and interface; Tisseron's dot, line, and circular tracing) that are in accordance with the idea of the symbolisation pre-verbal process as mapping.

Transitional therapists assume that this transitional state/experience can be 'relived' or 're-inhabited' under certain facilitating circumstances, to heal deficient sense of self constructs or to reconfirm them. Along with therapy, and under the generic term of 'cultural area of experience', Winnicott makes room for science and art (see also Aitken and Herman 1997, Bingley 2003). Indeed, Winnicott (2005), Anzieu (1981), and Tisseron (1987) have applied their transitional theories to art practice, extending their scope of relevance beyond therapy. They consider non-therapeutical practices as forms of transitional experiences, address artists' self-care strategies, and understand works of art as transitional constructs. Furthermore, Anzieu (1989, 7), by indicating 'a significant change in the nature of the suffering experienced by patients seeking for psychoanalytic help', understands borderline conditions and narcissistic suffering as significant conditions of the past decades in Western societies (see also Bondi 1999, who reflects upon the emergence of self-psychology). Hence, Anzieu (1989, 8) writes:

---

14 Indeed, an actual mapping process is depicted here: 'The baby has a concrete representation of this envelope which is provided for it by something of which it has frequent sensory experience (a sensory experience intermingled with phantasies) – its skin. It is these cutaneous phantasies which clothe [*habillent*] its nascent Ego with a figurative representation, admittedly imaginary, but which mobilizes, to borrow an expression to Paul Valéry, what is most profound in us, our surface' (Anzieu 1989, 60).

it seems to me a matter of the utmost urgency, in both psychological and social terms, for us to re-establish limits, restore some frontiers, mark out inhabitable, liveable territories for ourselves – limits and frontiers that will produce differentiation and at the same time allow exchanges between the regions (of the psyche, of knowledge, of society and humanity) thus delimited.

While Anzieu's agenda is primarily designed for transitional therapy, his remarks nonetheless provide interesting insights into geographers' fieldwork-as-withness. In line with Roussillon's (2008b) view on narcissism – that is, the subject effort to integrate at all costs the unachieved elements of the subjectivation process, and, consequently, his effort to bring back, through actual scenarios, the transitional processes –, I propose we understand fieldwork as a 'situation of transference'.

An agenda for developing transitionality in geography

We now go beyond elements in order to use transitionality as a theoretical framework of intelligibility for the nexus fieldwork/touch/sense of self and to elucidate a haptic regime of knowledge in geography, and consequently understand geography as transitional. This can be deployed in three different, interesting ways that show the general relevance of transitionality for geographical epistemology. I shall indicate the first two and concentrate on the third. All three work on the basis of the link between the sense of self and the experience of space/place through touch and attachment drive.

Firstly, the transitional theory of touch can be used to complement the post-phenomenological and non-representational theoretical basis of articulations about individuals' geographical haptic knowledges. It helps extend these articulations beyond specific somatic situations, as surveyed by Paterson (2009), and allows to encompass other significant – but, to my knowledge, unaddressed – dimensions of haptic knowledges, for example, the attachment to place or territorial/spatial identity. The French psychoanalyst Guillaumin (1975) contributed to a conference (organised by French geographers in Lyon) dedicated to the perception of landscape with a lecture that addressed the sense of landscape referring to Winnicott's notions of holding and use of the object, and that presented landscape as the matrix place/space of the self.

Secondly, the theory of transitional therapeutic practice can serve as a praxis model for the renewal of qualitative methodologies. This is precisely what Bingley (2003) did when she investigated adults' haptic sense of place/landscape through group excursions, sand modelling, and in-depth interviews designing her methodology on transitional references. By providing a facilitating environment that can be used/played by her research subjects, she explicitly adopted a transitionally based methodology. Her fieldwork methodology is close to Winnicott's notion of developing transitional phenomena within a transitional therapeutic framework as well as to French transitionality notions of touch and figurative processes of symbolisation. Geographers may draw on transitional practices (for instance, envelopments of sounds/words, art therapy, psychoanalytical drama, and so on) to extend their methodological tools for investigating geographical knowledges. In a non-representational perspective, geographers may resort to Winnicott's squiggle technique and Tisseron's elaboration of 'psychical imaging activity' to find ways to represent haptic knowledges, even in collaborative writings (Lloyd et al 2012) or map-makings (Burini 2008, Maulion 2008).

These theoretical and methodological perspectives on touch address the transitional link between sense of self and experience of place/space: fieldworkers work on the research respondents' sense of self to access their haptic knowledge of space/place. But, acknowledging the geographers' fieldwork non-academic psychoanalytical motives and strategies, we may consider conversely the way researchers make use of the spatial, relational, corporeal dimensions of their fieldwork practice (whether consciously or not) to bring their senses of self into play, and negotiate their self-identities through haptic experiences. From this perspective, and if we consider the results of this transitional process as personal benefits as well as research data, we complete the consideration for a haptic regime of knowledge. In other words, we come to examine epistemologically how field geographers' findings and constructs are sustained and shaped by transitional experiences on the basis of self-needs and by means of haptic processes, and how they represent them in geography. Therefore, thinking transitionally about fieldwork addresses what geographers' fieldwork haptic experiences are all about, but also questions what is the cognitive status of the haptically generated transitional data that geographers convert into knowledge objects.

Considering a Haptic Regime of Knowledge

Beyond considerations for haptic knowleges and methodologies, Guillaumin (1975) – the transitional psychoanalyst I mentioned earlier – and the French philosopher Wunenburger (1996) consider the transitional perspective of geography to be important. Wunenburger, who discusses the construction of knowledge in geography, brings together geographers' 'topophil drive' and their 'psychogeographical predisposition'. He puts forward the idea of a 'subject's interior atlas' that functions as a scheme for the geographical construct. Drawing on Guillaumin's, he resorts (briefly and somewhat metaphorically) to Winnicott's notions to explore geographical fieldwork as transitional experience. But, by ignoring French work on transitionality (the theory of touch), he finds no theoretical basis to sustain his assumptions: neither Anzieu's attachment drive, double anaclisis on skin and *entourage* through touch to articulate his idea of a 'topophil drive', nor Anzieu's skin-ego to articulate his assumed notion of a 'subject's interior atlas'. His ideas are encapsulated in a unique reference to the scopic regime of knowledge. When, for instance, he brings together geographers' use of maps and art of mapmaking and their fieldwork practice, he ponders them with a scopic frame of reference. Indeed, to question the transitional status of geography as a science, depends on the acknowledgement of its haptic regime of knowledge. Yet, with a completed transitional theory, we may go beyond Wunenburger's assumptions and elaborate the intelligibility of the nexus of fieldwork-as-withness/touch/self-identity so as to put forward the idea of a haptic regime of knowledge.

First, we may bring together a care-giving based therapy and a care-giving based fieldwork as well as, more generally, a fieldwork-as-withness model of research, insofar as touch and empathy are operators and processors of the ongoing processes of the therapeutic practice and the scientific practice. Second, we may underline the importance of spatial settings (clinical framework and field) and evolving spatial relationships (the clinical and the scientific fieldwork situations of holding/playing) in both processes. In particular, the ' 'all enveloping' ... spaces that are supportive' depicting the 'space of betweenness' (Duncan 1996, 6–7), the 'facilitating environment' (Bingley 2003) or the 'supportive atmosphere' (Kneale 2001) would stand on the holding side, while 'playing in the field' would stand on the side of the situation's use. Given the relational perspective of transitionality (see also Bondi 1999, 2003b) and of fieldwork, these elements may be submitted to the strict application of the principle of symmetry, so as to help us understand how the researched subjects and the researchers transitionally play their parts. Drawing on the tricky argument about touch in transitional therapy, and on the ways (empathy and envelopments, artistic activities and performances) transitional therapists have designed to circumvent the tactile while providing access to haptic data, we may gain new insights into some of the most popular qualitative methods: ethnography and in-depth one-to-one or group interviews, individual or participative group performances, and collaborative procedures of writing and drawing/mapmaking (see references above). This may also provide insights into other related issues, for example, the roles of interpreters and assistants (Hapke and Ayyankeril 2001) or sexuality as betweenness (Cupples 2002). Third, we may consider the role assigned to the activities of sketching, drawing, and mapmaking in geographical fieldwork (whether collaborative or not) and relate them to the psychical infra-verbal activities set out by Anzieu and Tisseron, on both the primary self-symbolisation process level and the therapeutic level of self-reconfirmation through haptic activities. This has two different implications. On the one hand, these activities of tracing – which occur in the course of fieldwork and which are part of the work – could be then understood as haptic-supportive activities of ongoing transitional processes of self-reconfirmation. On the other hand, their products may be recognised as mappings of skin-ego/scheme. This is a relevant basis for understanding the tricky issue of representing haptic knowleges (Paterson 2009). Finally, drawing on Wunenburger's idea of a 'subject's interior atlas', and given the spatial features with which Anzieu and Tisseron endow the skin-ego/scheme, we may make the following assumption: the skin-egos re-established within the course of the fieldwork practice and experience participate in the knowledge building, they function as schemes of the field-geographers' knowledge objects.

Conclusions

Firstly, I have sought to reveal the concealed haptic dimension of traditional fieldwork in geography, a dimension that still falls outside the scope of critical feminist examinations of geography. I then outlined the development of the haptic within fieldwork's qualitative reformation and discussed the contemporary examination of its epistemological basis. Hence, it has been argued that the sensuous and emotional experiences attached to the practice of fieldwork-as-withness entails an increasing questioning of researcher self-identity. I argue that this motive must be taken seriously as an epistemological matter and that the principle of symmetry should be applied to the fieldworker's search for self/care of self through haptic experiences. Hence, I maintain that we should turn qualitative research's transitional basis into a tool with which to examine a haptic regime of knowledge.

By combining Winnicott's spatial theory of the self and French transitional theory of touch, I discussed transitional psychoanalysis's spatial and haptic turns, i.e. the way in which the sense of the self and the experience of space/place are bound together through touch and attachment drive. I discussed the development of a care-giving based clinical setting in the theory of therapeutic practice, and the way it takes into account the spatial and the touch through various therapeutic techniques. I made specific points about Anzieu's, as well as Tisseron's, considerations for pre-verbal figurative processes, a mapping of the self through touch into body images and a haptic technique to support the generation of touch data and their psychical elaboration into self-image. Transitionality is then a relevant foundation of a haptic regime of knowledge, similar to Freudo-Lacanism for feminist poststructuralist geographers examining the scopic regime of knowledge of traditionally masculinist geography.

In the last section, I discussed the relevance of elements of transitional theory for an intelligible haptic regime of knowledge. I demonstrated how an examination of fieldwork-as-withness sustained by a transitional theory of touch, and that encompasses fieldwork practices, experiences, field-generated data, and field-based knowledge objects, allows one to address the fieldwork as well as the field-based geography as transitional (or transitionally informed). I reported on the works of geographers, philosophers, and psychoanalysts who have already partially articulated this matter in different ways. I indicated how this haptic regime, consolidated by the development of qualitative methodology, is commensurable with a more general identity or narcissistic condition at a social level – a borderline symptom identified by Anzieu on the basis of his therapeutic practice and to which he has dedicated his agenda.

Yet, a history of geographical fieldwork practice from the perspective of touch must be thought through and differentiated from the scopic regime of knowledge. This epistemological proposal must be documented further through empirical case studies[15] dealing with fieldworkers' practices, experiences, and knowledge objects or through the examination of a larger corpus of fieldwork practices' and experiences' accounts.

Acknowledgement

I am grateful to Ilse Evertse and Johan Emerson Grobler for their in-depth work of proof-reading.

References

Aitken, S.C. 2001a. Playing with children: Immediacy was their cry. *The Geographical Review*, 91(1/2), 496–508.

Aitken, S.C. 2001b. Shared lives: interviewing couples, playing with their children, in *Qualitative Methodologies for Geographers*, edited by M. Limb and C. Dwyer. London: Arnold, 73–86.

Aitken, S.C. and Herman, T. 1997. Gender, power and crib geography: transitional spaces and potential places. *Gender, Place and Culture*, 4(1), 63–88.

Anzieu, D. 1956. Intervention au *Discours* de Lacan, (26 Septembre 1956). *La Psychanalyse*, 1, 228–31.

Anzieu, D. 1974. Le moi-peau. *Nouvelle Revue de Psychanalyse*, 9, 195–203.

Anzieu, D. 1979. *Le Psychodrame Analytique chez L'enfant et L'adolescent*. Paris: PUF.

Anzieu, D. 1981. *Le Corps de L'œuvre. Essais Psychanalytiques sur le Travail Créateur*. Paris: Gallimard.

Anzieu, D. 1984. Au fond de soi, le toucher. *Revue Française de Psychanalyse*, 6, 1385–98.

Anzieu, D. 1989. *The Skin Ego* [translation Chris Turner]. New Haven and London: Yale University Press.

Anzieu, D. 1995[1985]. *Le Moi-Peau*. Paris: Dunod.

Anzieu, D. 1997. La demarche de l'analyse transitionnelle en psychanalyse individuelle avec des commentaires sur l'observation de Madame Oggi, in *Crise, Rupture et Dépassement*, edited by R. Kaës et al. Paris: Dunod, 186–221.

Balibar E., Cassin B. and Libera A. 2004. Sujet, in *Vocabulaire européen des philosophies*, edited by B. Cassin. Paris: Seuil / Le Robert, 1233–1254.

Berque, A. 1996. *Être Humains sur la Terre*. Paris: Gallimard.

---

15 After a break of several years, I recently resumed empirically investigating fieldwork processes with transitional perspectives. I have designed immersive methodologies by being with researchers in the field. My interests include focus group methodologies, ethnography and performances, as well as collaborative practices of drawing, mapping, and film-making.

Bingley, A. 2003. In here and out there: sensations between Self and landscape. *Social and Cultural Geography*, 4(3), 329–45.

Bleger, J. 1997. Psychanalyse du cadre analytique, in *Crise, Rupture et Dépassement*, edited by R. Kaës et al. Paris: Dunod, 257–76.

Bondi, L. 1999. Stages on journeys : Some remarks about human geography and psychotherapeutic practice. *The Professional Geographer*, 51(1), 11–24.

Bondi, L. 2003a. Meaning-making and its framing: a response to Stuart Oliver. *Social and Cultural Geography*, 4(3), 323–27.

Bondi, L. 2003b. Empathy and identification : Conceptual resources for feminist fieldwork. *ACME*, 2(1), 64–76.

Bondi, L. 2005. Making connections and thinking through emotions: between geography and psychotherapy. *Transactions of the Institute of British Geographers*, 30(4), 433–48.

Bowlby, J. 1969. *Attachment and Loss*. New York: Basic Books.

Burini, F. 2008. Cartographie participative et coopération environnementale en Afrique: le cas du village de Bossia (Niger). Paper to *Mapping Practices: Doing Fieldwork in Geography Conference*, France, Arras, 18–20 June 2008, <http://halshs.archives-ouvertes.fr/halshs-00389595/fr/>.

Chabert, C. 1996. *Didier Anzieu*. Paris: PUF.

Clancier, A. 1999. Une liberté de pensée. (Entretien avec D. Widlöcher), in *Le Paradoxe de Winnicott*, edited by A. Clancier and J. Kalmanovitch. Paris: In Press, 199–205.

Crang, M. 2002. Qualitative methods : the new orthodoxy. *Progress in Human Geography*, 26(5), 647–55.

Crang, M. 2003. Qualitative methods: touchy, feely, look-see? *Progress in Human Geography*, 27(4), 494–504.

Crang, M. 2005. Qualitative methods: there is nothing outside the text? *Progress in Human Geography*, 29(2), 225–33.

Cupa, D. 2000. La pulsion d'attachement selon Didier Anzieu et la relation de tendresse, in *L'attachement. Perspectives Actuelles*, edited by D. Cupa. Paris: EDK, 97–119.

Cupples, J. 2002. The field as a landscape of desire: sex and sexuality in geographical fieldwork. *Area*, 34(4), 382–90.

Dardel, E. 1952. *L'homme et la Terre*. Paris : PUF.

Delion, P. 1998. *Le packing avec les enfants autistes et psychotiques*. Ramonville Saint-Ange: Eres.

Delourme, A. 1997. *La Distance Intime. Tendresse et Relation d'aide*. Paris: Desclée de Brouwer.

Duncan, N. 1996. Introduction. (Re)placings, in *Body Space: Destabilizing Geographies of Gender and Sexuality*, edited by N. Duncan. London: Routledge, 1–10.

Golse, B. 1999. *Du Corps à la Pensée*. Paris: PUF.

Green, A. 1999. Winnicott et le modèle du cadre (entretien), in *Le Paradoxe de Winnicott*, edited by A. Clancier and J. Kalmanovitch. Paris: In Press, 171–78.

Green, A. 2000. Le moi et la théorie du contact, in *Les Voies de la Psyché. Hommage à Didier Anzieu*, edited by R. Kaës et al. Paris: Dunod, 217–26.

Guillaumin, J. 1975. Le paysage dans le regard d'un psychanalyste, rencontre avec les géographes. *Cahiers du Centre de Recherche sur l'Environnement Géographique et Social*, 3, 12–35.

Hapke, H.M. and Ayyankeril, D. 2001. Of 'loose' women and 'guides', or relationships in the field. *The Geographical Review*, 91(1/2), 342–52.

Kaës, R. 1997[1979]. Introduction à la pyschanalyse transitionnelle, in *Crise, Rupture et Dépassement*, edited by R. Kaës et al. Paris: Dunod, 1–83.

Kaës, R., Missenard, A., Kaspi, A., Anzieu, D., Guillaumin, J., Bleger, J. and Jaques, E. 1997[1979]. *Crise, Rupture et Dépassement*. Paris: Dunod.

Katz, C. 1994. Playing the field: Questions of fieldwork in geography. *The Professional Geographer*, 46(1), 67–72.

Kaspi, A. 1997[1979]. L'histoire de la cure psychanalytique de Madame Oggi, in *Crise, Rupture et Dépassement*, edited by R. Kaës et al. Paris: Dunod, 149–85.

Khan, M.R. 1971. Une certaine intimité, in *La Consultation Thérapeutique et L'enfant*, edited by D.W. Winnicott. Paris: Gallimard.

Kneale, J. 2001. Working with groups, in *Qualitative Methodologies for Geographers*, edited by M. Limb and C. Dwyer. London: Arnold, 136–50.

Latour, B. and Woolgar, S. 1996. *La vie de laboratoire. La production des faits scientifiques*. Paris: La Découverte.

Lawrence, K. 2010. Hanging from knowledge: Vertical dance as spatial fieldwork. *Performance Research*, 15(4), 49–58.

Lecourt, E. 1987. L'enveloppe musicale, in *Les Enveloppes Psychiques*, edited by D. Anzieu. Paris: Dunod, 199–222.

Limb, M. and Dwyer, C. 2001. Introduction: Doing qualitative research in geography, in *Qualitative Methodologies for Geographers*, edited by M. Limb and C. Dwyer. London: Arnold, 1–20.

Lloyd, K., Wright, S., Suchet-Pearson, S., Burarrwanga, L.L. and Hodge, P. 2012. Weaving and working together: collaborative fieldwork narratives in North East Arnhem Land, Australia, in *Terrains de Je. (Du) Sujet (au) géographique*, edited by A. Volvey, Y. Calbérac and M. Houssay-Holzschuch. *Annales de Géographie*, 5(687), 110-118.

Longhurst, R., Ho, E. and Johnston, L. 2008. Using 'the body' as 'instrument of research'. *Area*, 40(2), 208–17.

Maulion, H. 2008. Narrer l'expérience intime du terrain. Paper to *Mapping Practices: Doing Fieldwork in Geography Conference*, France, Arras, 18–20 June 2008, <http://halshs.archives-ouvertes.fr/halshs-00357433/fr/>.

Myers, G.A. 2001. Protecting privacy in foreign fields. *The Geographical Review*, 91(1/2), 192–200.

Nast, H.J. (ed.) 1994a. Women in the Field (theme issue). *The Professional Geographer,* 46(1).

Nast, H.J. 1994b. Opening remarks on 'Women in the Field'. *The Professional Geographer*, 46(1), 54–66.

Nast, H.J. 1998. The body as 'place'. Reflexivity and fieldwork in Kano, Nigeria, in *Places Through the Body*, edited by H.J. Nast and S. Pile. London: Routledge, 93–116.

Nast, H.J. and Kobayashi, A. 1996. Re-corporealizing vision, in *Body Space: Destabilizing Geographies of Gender and Sexuality*, edited by N. Duncan. London: Routledge, 75–93.

Parr, H. 2001. Feeling, reading and making bodies in space. *The Geographical Review*, 91(1/2), 158–67.

Paterson, M. 2009. Haptic geographies: ethnography, haptic knowledges and sensuous dispositions. *Progress in Human Geography*, 33(6), 766–88.

Pourtier, R. 1991. Derrière le terrain, l'Etat', in *Histoires de Geographes*, edited by C. Blanc-Pamard. Paris: Éditions du CNRS, 91–102.

Powell, R.C. 2002. The sirens' voices. Field practices and dialogue in geography. *Area*, 34(3), 261–72.

Prayez, P. 1994. *Le toucher en psychothérapie*. Paris: Hommes et Perspectives/ épi-DDB.

Punch, S. 2001. Multiple methods and research relations with children in rural Bolivia, in *Qualitative Methodologies for Geographers*, edited by M. Limb and C. Dwyer. London: Arnold, 165–80.

Rogers, C. 1957. The necessary and sufficient conditions of therapeutic personality change. *Journal of Counselling Psychology*, 21, 95–103.

Rose, G. 1993. *Feminism and Geography: The Limits of Geographical Knowledge*. Minneapolis: University of Minneapolis Press.

Rose, G. 1996. As if the mirrors has bled, in *Body Space: Destabilizing Geographies of Gender and Sexuality*, edited by N. Duncan. London: Routledge, 56–74.

Rose, G. 1997. Situating knowledges: positionality, reflexivities and other tactics. *Progress in Human Geography*, 21(3), 305–320.

Roussillon, R. 1995. *Logiques et Archéologiques du Cadre Analytique*. Paris: PUF.

Roussillon, R. 1999. Actualité de Winnicott, in *Le Paradoxe de Winnicott*, edited by A. Clancier and J. Kalmanovitch. Paris: In Press, 9–26.

Roussillon, R. 2008a. *Le Jeu et L'entre-je(u)*. Paris: PUF.

Roussillon, R. 2008b. Corps et actes messagers (introduction), in *Corps, acte et symbolisation. Psychanalyse aux frontiers*, edited by B. Chouvier and R. Roussillon. Bruxelles: de Boeck, 23–37.

Sharp, J. 2005. Geography and gender: feminist methodologies in collaboration and in the field. *Progress in Human Geography*, 29(3), 304–9.

Sotelo, L.C. 2010. Looking backwards to walk forward: Walking, collective memory and the site of the intercultural in site-specific performance. *Performance Research*, 15(4), 59–69.

Sparke, M. 1996. Displacing the field in fieldwork. Masculinity metaphor and space, in *Body Space: Destabilizing Geographies of Gender and Sexuality*, edited by N. Duncan. London: Routledge, 212–33.

Tisseron, S. 1987. *Psychanalyse de la Bande Dessinée*. Paris: PUF.

Tisseron, S. 1995. *Psychanalyse de L'image*. Paris: Dunod.

Tisseron, S. 1996. *Le Bonheur dans L'image*. Paris: Synthélabo Ed.

Volvey, A. 2000. L'espace vu du corps, in *Logiques de L'espace, Esprit des Lieux. Géographies à Cerisy,* edited by J. Lévy and M. Lussault. Paris: Belin, 319–32.

Volvey, A. 2003. *Art et Spatialités D'après L'œuvre D'art in Situ Outdoors de Christo et Jeanne-Claude.* Thèse de Doctorat, Université Paris I – Sorbonne.

Volvey, A. 2004. Übergänglichkeit: ein neuer Ansatz für di epistemologie der geographie. *Geographische Zeitschrift*, 92(3), 170–84.

Volvey, A. 2010. Spatialités du land art à travers l'œuvre de Christo et Jeanne-Claude', in *Activité Artistique et Spatialité*, edited by A. Boissière, V. Fabbri and A. Volvey. Paris: L'Harmattan, 91–134.

Volvey, A. forthcoming. Le terrain transitionellement: une transdisciplinarité entre géographie, art et psychanalyse, in *1970-2010: Les sciences de l'homme en débat*, Actes du colloque sur les transformations des sciences humaines et sociales depuis 1970, Nanterre : Presses Universitaires de Paris Ouest – PUF.

Volvey, A., Calbérac, Y. and Houssay-Holzschuch, M. 2012. Terrains de Je. (Du) Sujet (au) géographique (introduction), in *Terrains de Je. (Du) Sujet (au) géographique*, edited by A. Volvey, Y. Calbérac and M. Houssay-Holzschuch. *Annales de Géographie*, 5(687), 3–40.

Volvey, A. and Houssay-Holzschuch, M. 2007. La rue comme palette. *La Pietà sud–africaine*, Soweto/Warwick, mai 2002, Ernest Pignon-Ernest, in *Spatialités de L'art*, edited by A. Volvey. T.I.G.R. 33(129-130), 145–74, <http://halshs. archives-ouvertes.fr/halshs-00426890/fr/>.

Winnicott, D.W. 2005[1971]. *Playing and Reality*. London: Routledge.

Wunenburger, J-J. 1996. Imagination géographique et psycho-géographie, in *Lire L'espace*, edited by J. Poirier and J.-J. Wunenburger. Bruxelles: Ousia, 399–414.

Chapter 6

Guiding Visually Impaired Walking Groups: Intercorporeal Experience and Ethical Sensibilities

Hannah Macpherson

Introduction

As a sighted guide, traversing the Lake District hills with a visually-impaired companion, I feel their hand on my elbow. That hand – its grip and changing orientation – helps me to feel the walker's movements behind me and I adjust my pace accordingly. I also feel worthwhile – I am traversing a hillside, not just for the exercise or sheer joy of the view, but to be someone else's eyes and for the joy of an immediate and required connection with somebody else. I have a clear purpose, a seemingly worthy role and a grateful companion – a companion who has been partly incorporated into my own 'body schema' as I am into theirs. This practice of guiding requires learning the habit of 'sensing for two' and moving 'as one'. (Macpherson, edited extract from Reflective Field Notes, 2006.)

Being a sighted guide for people with blindness and visual impairment involves a range of tactile-kinaesthetic connections between two different people and a habitual practice of 'sensing for two' and 'moving as one'. This chapter explores how these experiences are significant for participants in the guiding relationship and for attempts to understand the relations between embodiment, touch and ethics (Varela 1999, Weis 1999, Diprose 2002, McCormack 2003). The chapter is developed from qualitative research material generated while acting as a sighted guide with specialist visually impaired walking groups who visit the British countryside (Macpherson 2008, 2009a and b). Sensuous ethnographic observation is used alongside interview material and photography to give an account of the interconnected, haptic and habitual qualities of the guide-walker relationship. The chapter reflects critically on who needs who in such 'worthy volunteer' situations and what exactly is being 'given' by the bodies of volunteer and visually impaired walker.

Specifically the chapter focuses on the intercorporeal spaces that emerge within the guide-walker relationship. Such intercorporeal spaces that occur between bodies have had a tendency to be overlooked by research which focuses on the individual body (cf. Thrift 1997 and 2007) and by geographic research which focuses on visual impairment as simply a 'way finding problem' or as an individual lived experience (Hill 1985, Gollege 1992, Jacobson and Kitchin 1997, Kitchin et al 1997). Therefore the concept of intercorporeality is used here 'to emphasise that the experience of being embodied is never a private affair, but is always already mediated by our continual interactions with other human and nonhuman bodies' (Weis 1999, 5). Building on an intercorporeal approach to understanding embodiment it is revealed that the guide-walker relation is *not* a matter of one worthy volunteer *taking* a grateful blind person through an unfamiliar rural environment – a pre-conception of many observers and first time guides. Rather, it is shown that a successful guide-walker relationship is more akin to the *kinetic synchrony* achieved between partners in a dance – a set of pre-reflective movements founded upon a coupling of their body-schemas that require only limited verbal accompaniment (Olsezweski 2008).

This touch of two bodies in movement transforms participant's sensibilities for it involves a 'sensing for two' and moving 'as one'. This is important, not only because it is a novel empirical observation about people with visual impairments tactile-corporeal relationship to sighted guides, but because it has implications for how geographers, including researchers of (dis)ability, approach research on 'the body'. Specifically it points to a need to take seriously *the coupled, intercorporeal qualities of bodies* that touch and move together, including how they open up and enact particular forms of corporeal-ethical spaces.

The chapter is structured as follows: first, I review previous geographic research on visual impairment and suggest a need to move research beyond a focus on individual way finding behaviour or socio-spatial experience. Secondly, I describe the methods and movements involved in guiding and show how this practice requires both guide and walker to habitually 'sense for two' and 'move as one'. Thirdly, I note that for synchronous movement to be achieved the visually impaired walker must sacrifice their independent ability to way-find and bestow upon the guide a 'gift of trust' in order to move through an unfamiliar rural environment at walking pace. Fourth, I argue that acknowledging this 'gift of trust' is significant because it is more often the volunteer guide who symbolically occupies the 'worthy role' in the guide-walker relationship.

The gift of trust that visually impaired walkers bestow upon their guides also draws attention to a range of forms of 'corporeal generosity' (Diprose 2002) that are at work within the guide-walker relationship, where this generosity is understood as 'being given to others without deliberation in a field of intercorporeality' (p. 4). Therefore in the final section I explore in more depth how successful guiding movement involves a coupling of the 'body schemas' (Merleau-Ponty 1962) of walker and guide and a form of corporeal generosity. In so doing I make a contribution to a renewed geography of disability which recognises the intersecting and interdependent nature of (dis)abled and able bodies (Weiss 1999, Crooks 2010, Power 2010) and a contribution to non-representational concerns with the walking body, the habitual and the taken for granted dimensions of experience (Thrift 1997 and 2007, Harrison 2000, McCormack 2003, Harrison 2008, Wylie 2006, Bissell 2010).

Researching Experiences of Visual-impairment

Social scientific and geographic research involving people with visual impairments has tended to be dominated by either, practical research concerned with how people with visual impairments can navigate urban environments effectively (Gollege 1992, Jacobson and Kitchin 1997, Kitchin et al 1997) or are concerned with social representations, attitudes and stereotypes that are encountered by these people (Butler and Bowlby 1997, Watson 2003). In such studies there tends to be an assumption that the experience of visual impairment is one of social and spatial disadvantage (cf. Allen 2004). There have also been some attempts to explore the lived experience of visual impairment as an active 'body-in-space encounter' (Allen 2004). Such an approach includes the work of Hetherington (2002 and 2003) in his exploration of visually impaired people's non-representational experiences of museum objects, along with a range of phenomenological studies (e.g., Hill 1985, Cook 1992) which take inspiration from Merleau-Ponty (1962) and his specific example of the extension of body-schema through the cane of a person with blindness.

My own work can be located within these latter, phenomenogically inspired approaches. Yet I show how people with blindness and visual impairment are not only immersed in the materiality of the world but also in the materiality and movements of other people's bodies. This intercorporeal approach to understanding visual impairment is distinct from the popular genre of autobiographical writing which insists that blindness creates its own 'world' or 'way of knowing' distinct from those who are sighted (Hull 1990 and 2001, Kleege 1998). For example, Professor of theology John Hull (2001, 23) has suggested that 'blindness is something which creates its own worlds'. However, I attend to how people with blindness are also involved in a *co-emergent world* involving joint ways in which they orientate and attune themselves. For as deaf-blind author Helen Keller (1908, 58) put it over a century ago:

> It might seem that the five senses would work intelligently together only when resident in the same body. Yet when two or three are left unaided, they reach out for their compliments in another body, and find that they yoke easily with the borrowed team. When my hand aches from overtouching, I find relief in the sight of another. When my mind lags, wearied with the strain of forcing out thoughts about dark, musicless, colorless, detached substance, it recovers its elasticity as soon as I resort to the powers of another mind which commands light, harmony, color.

Keller (1908) draws attention to the way in which she sometimes relies on other's sensory experience explaining that 'they yoke easily with the borrowed team'. Blind sociologist Rod Michalko (1999) communicates this concern with shared corporealities in his autobiographical book, *The Two in One: Walking with Smokie, Walking with Blindness*. He reflects on his experience of navigating the environment with Smokie his dog and on departing guide dog training school, writing that: 'I was now ready to leave; I had arrived at the school as simply "me" but I was leaving as "we," as part of a dog guide team.' (p. 82). The 'two in one' for Michalko was the conjoining lives of a person with blindness and their dog guide. The 'two in one' relation that concerns me here is the felt relation between a person with a sight impairment and their human guide.

Sensuous Ethnography and the Work of the Sighted Guide

The methodological approach to the research discussed in this chapter has involved using my own body as an 'instrument of research' (Crang 2003) in a form of reflexive, embodied ethnography acting as a sighted guide. This form of 'sensuous scholarship' advocated by Paterson (2009) and other researchers interested in accessing the 'non representational' (Thrift 1997) or 'more-than-representational' (Lorimer 2005) requires bodily immersion in a context, an openness to the experiences and feelings of others and critical reflection on those experiences and the transformations undergone by the researcher. Here there is a concern with what is felt and done as well as what is seen and said. For example, Stoller (1997) uses such an approach to consider how the Songhay people of Nigeria experience their world and defines sensuous ethnography in the following manner:

> Sensuous ethnography, of course, creates a set of instabilities for the ethnographer.
> To accept sensuousness in scholarship is to eject the conceit of control in which
> mind and body, self and other are considered separate. (Stoller 1997, xvii)

Stoller's insight into what a sensuous ethnography might require resonates with my own experiences as an researcher and sighted guide where I attempted to 'experience with' visually impaired walkers, rather than simply observe. This involved accepting the transformative (rather than simply observational) nature of social research (Whatmore 2003) and reflecting on what it was like to go through the process of learning to be a guide and to end up 'seeing and feeling for two' and 'moving as one'. Therefore, in this section I will firstly outline some of the methods and body practices that are involved in learning to be a competent guide and then develop a set of reflections on the motivations, rewards and sensibilities of being a guide.

Sighted guides help people with visual impairments explore areas outside of their known routes. Throughout the United Kingdom volunteer guides are recruited by visually impaired charities and adverts for guides can regularly be seen on local volunteer recruitment websites. This research focused on participants of a Sheffield based visually impaired walking group and the visually impaired holiday charity *Vitalise*. My research was initially motivated by a 'politics of representation' and a concern with representing under-represented visitors to the countryside. Therefore, I was particularly interested in the visually impaired members of these

Figure 6.1   The 'c' grip method of guiding
Source: author photograph. Consent granted by those individuals featured

groups because they visited two landscapes of national significance in Britain – the Lake District and the Peak District. However, since beginning the research I also became increasingly interested in the relationship established between walker, guide and landscape; and the affective, corporeal components of these relations (cf. Macpherson 2008 and 2009a).

There are number of styles and methods of guiding people with visual impairments. By far the most common method is the 'c' grip (Figures 6.1 and 6.2). Here the person with a visual-impairment will hold onto the guide's arm above the elbow with a 'c' grip in order to navigate their way through a new environment. This method allows the person with a visual impairment to stay one step behind the guide. Through their movements and their verbal instructions the guide can warn the person who is being guided of any forthcoming hazards. Thus, the 'c' grip itself as a direct tactile connection is also part of a wider body practice of sensations, muscular tensions, movements and balance that take account of the needs of two people to navigate the terrain in a safe and effective manner. The 'c' grip method is widely regarded by visual impairment rehabilitation professionals in the United

Figure 6.2   The 'c' grip method of guiding

Source: author photograph. Consent granted by those individuals featured

Kingdom as a preferable method to linking arms because it allows for a verbal and physical warning of forthcoming hazards. It is also thought to allow for a more 'impersonal' (less intimate) form of touch (interview with rehabilitation officer, Jade[1], June 2005). Yet the 'c' grip may in fact be more intimate than its description implies, because it is the start of a wider sensate body practice required for synchrony to be achieved.

Interestingly, the preferred methods for guiding vary between countries and this reveals something of the differing cultural attitudes towards forms of guiding touch; for example, in Greece it is more common to link arms (Charidi 2009). In Britain methods for guiding in the countryside also vary; the linking arms was found to be a common method amongst friends and partners and was preferred by some participants because this method less obviously 'marked them out' as having a visual impairment (Figure 6.3). Linking arms could also provide a strong and

---

1   Pseudonyms have been used here to protect the identity of participants. The people featured in the photograph have given full consent for their use and to ensure anonymity they are not the sources of the interviewee quotes in this chapter.

Figure 6.3   Methods of guiding visually impaired and blind walkers in the countryside – linking arms

Source: author photograph. Consent granted by those individuals featured

reassuring level of physical support from the sighted guide who could be leant on if necessary. Touching the back of the rucksack or holding onto a strap on the back of the rucksack were also common methods of guiding (Figure 6.4). These methods were useful for narrow paths, for steep descents or for when the walker wanted to feel a greater sense of freedom. Handholding was also a useful tactic for tricky areas or a method of guiding used amongst friends and partners (Figure 6.5).

Each method of guiding and each encounter with different terrain required the walker and guide to respond to each other's movements in quite specific ways. Visually impaired walkers who were able to balance well and did not need a direct connection to someone's arm were able to make good use of the strap on the back

Figure 6.4   Methods of guiding visually impaired and blind walkers in the countryside – touching the back of rucksack

Source: author photograph. Consent granted by those individuals featured

of the rucksack to guide them. When holding a strap their movements were often freer and more independent of the guide than when using the 'c' grip method. In these cases their response and interpretation of forthcoming hazards and changes in terrain was directed toward both the guides movements and the material landscape, which they would navigate through their feet, their tactile-kinaesthetic senses and through the use of a walking pole or white cane as an extension of their 'body schema' (Merleau-Ponty 1962, 143). In fact the lightness of the connection between walker and guide when using a strap would at times mean it would take the guide some time to realise if the walker had dropped the strap and lost them completely. This was certainly a disadvantage of this method.

In contrast the 'c' grip method of guiding gives the visually impaired walker a direct and continuous felt connection to their guide and the walker could read the landscape and respond to the changes in terrain through the movement of the guide and their own tactile-muscular and kinaesthetic senses. As a guide I found it to be my preferred method because I could literally feel the location and movements of the walker I was guiding and I could co-ordinate my movements with them in order to move 'as one' across the hillsides. For example, subtle movements of hand position and grip combined with a proprioceptive sense of my own and the other's bodily position would make me aware that the person I was leading had fallen too

Figure 6.5   Methods of guiding visually impaired and blind walkers in the countryside – direct handholding

Source: author photograph. Consent granted by those individuals featured

far behind me and I could adjust my pace accordingly. This required a 'response-ability' of walker and guide to be directed toward each other's movements and the terrain. In general only limited verbal interpretation of the terrain was required when using the 'c' grip, because the walker could feel the movements of the guide and respond accordingly. Flatter terrain meant that walker and guide could synchronise their movements more easily and conversation about a range of topics could flow (if so desired). In contrast traversing steep or uneven terrain required a sort of constant somatic vigilance that often precluded any conversation; instead walker and guide would concentrate on each other and on changes in the terrain.

At best, the co-ordinated dual movement across the hillsides gave me a rewarding sense of silent camaraderie with the person I was leading. It was a quiet, practical means of going, founded in our ability to respond to a situation. This has been referred to in previous literature on embodied ethics as a form of 'ethical know-how' (Varela 1999) or 'response-ability' (Caputo 2003). For in these cases the response to the other person involved a form of 'flow' – a responding to the other in movement, where response-ability is understood 'to be a matter of settling into the singular demands of a situation and responding, a matter of being taken hold of by the situation and facing up to what it demands of us' (Caputo 2003, 71). Thus guiding and being guided involved a pre-reflective, habitual movement together comparable to the synchronous movements achieved between dance partners. This dual movement has been elaborated upon by Charidi, a Greek anthropologist who has been currently researching the practice of being a sighted guide in Athens. In her auto-ethnographic writing she describes the guide-walker movement as follows:

> In general, both the blind and the sighted try to move in a co-ordinated way, let's say schematically 'as one body'. This draws the practice of guidance out of its authoritative use of 'showing the way' and puts together blind and sighted in the world of sight's absence. 'You have to think you are taller and double in size', Stefanos explained to me, during the first times I guided him. On the one hand, I had to move in the space 'like' him and see 'instead' of him. On the other hand, he had to 'lean on' my eyes and perceive through my body what I see. Thus, moving 'as one body' entails a mutual motion towards the other: the blind has to attend to his guide's body and the sighted has to encompass in her own body that of the guided. (Charidi 2009, 4)

Charidi discussed how Stefanos teaches her to guide, by telling her to move through a space imagining she is twice the size. For her, this involves a 'mutual motion toward the other'. I think such a conception of guiding is important because it acknowledges the mutual ties and reciprocal (rather than authoritative) nature of the guide-walker relationship. Furthermore, it begins to show the way in which guiding is important not only because it can potentially make a volunteer guide feel better about themselves (and the recipient of the guiding progresses to somewhere they could not have gone otherwise), but also because it makes the guide and walker feel the other's immediate needs. In the setting of the Lake District and Peak District this dual movement did not always run smoothly though. Some walkers would lean heavily on their guides upsetting their guide's sense of balance and making it difficult to respond and co-ordinate guide-walker movement effectively. Also, while guides would try and navigate the landscape for two at times this could become difficult – steep descents or rutted paths were particular challenges to the smooth coordinated movement of walker and guide.

Mistakes were sometimes made by the guide. For example, I once forgot that my companion was taller than me and we crashed into some overhanging branches. Such mistakes are interesting, firstly, because they highlight how the majority of the time guide and walker manage to respond to each other effectively and move smoothly through the landscape in a synchronised, response-able manner. Secondly, they are significant events because they highlight the level of trust that is bestowed upon the guides in these settings. I will attend to this latter issue regarding the 'gift of trust' here, before reflecting further on the accomplishment of smooth dual movement.

The Gift of Trust

Most of the visually impaired walkers that attend the trips are accomplished at navigating their local environments without the use of a sighted guide, typically exploiting a combination of memory, habit, a cane or a guide dog. However, visually impaired walkers give up their achieved, independent ability to way-find in order to move through areas of challenging terrain with a sighted guide. The human guide was necessary because guide dogs are trained to navigate paved, urban environments and information gained from walking cane, residual sight, tactile and audio clues did not tend to be enough to navigate unfamiliar rural spaces successfully with partial sight. Thus, walkers needed to hold the guide in some fashion and trust the guide's movements in order to move at an average walking pace through rural terrain. In this way, by entering an unfamiliar rural environment the visually impaired walker is placing themselves in a position of vulnerability. In such spaces I think they bestow upon the guide a sort of 'gift of trust'. The following quote from congenitally blind walker Jim begins to illustrate this gift of trust and the vulnerability felt by the visually impaired walker in the countryside:

> I think I feel more of a sense of freedom in the city in an area I know – admittedly they are small – than I do in the country, where I am totally dependent on a sighted guide... If I lost my grip on the guide's rucksack or sleeve I would immediately have to stop and call to my guide and say, could you stop? I have lost your sleeve, because that in a sense is your lifeline, if you lose your sighted guide you could be in trouble. (Jim, congenitally blind, 25–35 years old; interview in August 2004.)

Jim refers to the guides sleeve as is 'lifeline' and he is dependent on the guide to navigate the environment for them both. He trusts the guide in order to enjoy the other symbolic, physical and social rewards that walking in the countryside can afford. This dependence of the walker on the guide is a largely enjoyable (rather than burdensome) experience for the volunteer guide because it bestows upon them a feeling of responsibility and a worthwhile role. This 'gift of trust' is also found to occur in urban as well as rural settings, for example, in the following extract we can see how Ellen, who has macular degeneration, gives her visual impairment as a gift to a helper whose assistance she does not really need:

> Sometimes I will be standing waiting to cross the road and someone will come up to me and say, 'are you wanting to cross the road?' And I will think yeah I would be crossing the road if you just shut up and let me listen to the traffic! Do you know what I mean? (laughs) But being courteous and sometimes allowing them to help me is as much for them as for me...It is a fine line... and you know sometimes I am just too damn lazy and I just think, oh I will let some body else do that (both laugh) ... sometimes some little person will come up to me and say can I help

you? and well I do think, ok you don't get much chance to help people, go on then.
(Ellen, macular degeneration, 55–65 years old; interview in August 2004.)

Ellen explained to me how sometimes she would just 'let' someone guide her even though she did not necessarily require guiding at that point. Thus she offers her visually impaired status as a charitable gift toward someone who would otherwise not 'get much chance to help people'. Ellen, like Paul quoted next and Jim in the earlier extracts, lets herself be guided by people who are sighted. This is significant because it is more common to recognise that the sighted guides are 'giving up' something as volunteers in order to 'help' the visually impaired walkers. For example, at the disabled holiday charity *Vitalise* (read 'vital-eyes') sighted guides pay a subsidised trip fee (in order to recruit adequate numbers) and people with visual impairments pay the full cost of the holiday. *Vitalise* advertise their week long walking trips to sighted guides with the following text on their website:

> Would you like to enjoy your holiday knowing that you've also helped someone else get the most out of theirs?

Sighted guides are an essential part of Vitalise Holidays. You could make the difference between a visually impaired person having a much-needed break or no holiday at all. Not only that, but as a sighted guide you will be adding a whole new dimension to the holiday for our visually impaired customers, whether you're describing the Sistine Chapel, navigating the Lochs of Scotland or narrating the thrilling last moments of a Cup Final. You could even find that you're looking at things in a new way yourself. So becoming a sighted guide enhances the holiday of our visually impaired customers and is a unique and incredibly rewarding experience for the guide. It can sometimes be demanding, but it's also great fun. (Vitalise 2007)

The advert for guides mentions that the sighted guides are an 'essential part' of the holiday and that this will be 'rewarding experience'. However, the rewards on offer remain somewhat undefined. Participant explanations would draw upon a range of charitable discourses, ethical principles and personal stories in order to explain some of the motivations and rewards of guiding. Some participants drew on broader charitable principles around 'doing good' for others to explain their guiding practice but unfortunately these could end up reproducing a 'tragic' or 'pity' narrative of the experience of visual impairment. For example, Angela states: 'I wanted to help somebody who was in a worse off position than myself, because I feel so privileged in day to day life' (Angela, sighted guide, 45–55 years old; interview in August 2005). While Julie describes her motives as follows: 'I wanted to give a visually impaired person an opportunity they won't get in their everyday life…because one is so conscious of the fact that so many visually impaired people go away and they don't get to go walking again until they come back on one of these holidays, which is just so awfully sad.' (Julie, sighted guide, 55–65 years old; interview in July 2005.)

This kind of self-stated justifications for volunteering as a guide risk perpetuating an outdated 'tragedy model' (Swain et al 2004) of disability because they rely on a distinction being made between the volunteer who is supposedly 'better off' and the 'pitiable' visually impaired person who requires their charitable act. However, it is important though to question who needs who in such volunteer situations and what exactly is being given. Allaharyi (2000) in her study of volunteers in Sacramento, California, refers to a problem of 'moral-selving' where conceptualising of volunteering as worthy runs the risk of denying the obvious 'recipient' agency and elides the possibility of mutual ties. This produces an 'altruistic paradox'. So when a guiding volunteer such as Julie says she wants to '*give* a visually impaired person an opportunity they won't get in their everyday life' or when Angela states that she 'wanted to help somebody who was in a worse off position' they maybe unwittingly implicated in reinforcing a lower status position for the person being helped (see also discussion in Douglas 1990, ix).

In recognising the issue of 'moral-selving' Sennett (2004, 140) asks us to acknowledge that volunteering 'for' others may simply 'serve the more personal need to affirm something in ourselves'. I would agree that practising as a sighted guide may indeed involve an element of this 'worthy self-affirmation'. However, it is also important to note that practising as a guide presents people with an opportunity to experience a transformation. This is because in walking together with someone as a sighted guide, they embody the person's sight impairment as they 'sense for two' and 'move as one'. Thus guiding rather than simply being an act of self realisation may also force an encounter with the 'ambiguity of existence' (Diprose 2002, 90), because the guide experiences something in themselves that they have not necessarily encountered before – a coupled body schema and a capacity for habitual synchronous movement.

Thus the appeal and 'rewards' of this form of volunteering may not be located in simple 'self affirmation' but rather in the displacement and disruption of a coherent sense of self; the challenge from feeling different. In the following section I, therefore, explore in more detail the dual movement of walker and guide, and shows how guiding and being guided involves a pre-reflective 'coupling of body schemas'. This coupling is shown to occur as the guide learns to be a competent guide and as the walker learns to trust and follow the guide's movements.

Learning to be a Competent Guide

Prior to meeting with the walking groups I had read some of the literature on how to guide a person with blindness and attended a training course for volunteers working with blind and visually impaired people delivered by the Newcastle Society for the Blind. I had hoped this preparation would enable me to act as a competent sighted guide from the outset, but, initially acting as a sighted guide did not 'come naturally'. I noticed at the start that I was a rather hesitant, clumsy and unsure guide in comparison to some of my more accomplished peers. As a new guide 'looking for two' and 'moving as one' required conscious mental effort. This mental concentration of acting as a guide changed over time and during my third outing into the countryside with visually impaired walkers I gradually began to find myself feeling more at ease as a guide – thinking less and utilising a range of skills and physical dispositions which were suitable to the sightless group dynamic. For example, I learnt to utilise touch to introduce myself as well as sound – a verbal signal and a hand on the arm to make a connection (because no eye contact was available). I would also instinctively shift my arm back as we were getting off a bus so that the person behind me could hold onto it if they needed to.

My attention was drawn to the habitual nature of these body practices when, after spending a day with the visually impaired group, I would still find myself navigating the terrain by allowing for the width of two people on a path and watching the terrain for low branches. This meant I was absorbing the person being guided into my own habits of practice and 'body schema' (Merleau-Ponty 1962). This is a sense of bodily awareness is elaborated upon by Merleau-Ponty (1962, 165) when he writes of the extension of body schema through the cane of the person with blindness:

> The blind man's stick has ceased to be an object for him, and is no longer perceived for itself; its point has become an area of sensitivity, extending the scope and active radius of touch, and providing a parallel to sight. In the exploration of things, the length of the stick does not enter expressly as a middle term: the blind man is rather aware of it through the position of objects than of the position of objects through it.

Here Merleau-Ponty's example of the 'blind man's stick' shows how a sense of bodily awareness is not necessarily limited to the outer surface of the skin but can become extended through the cane. It is also possible to apply this conception of an extended body schema to other aspects of bodily and environmental experience. For example, Allen (2004) in his research into the way finding behaviour of visually impaired children, shows how they can incorporate fixed elements of their home space and neighbourhood into an 'extended corporeal schema' (p. 734). This enables visually-impaired children to navigate in an effective, habitual manner that is not reliant on obvious cognitive maps. Similarly, guiding and being guided involves an extended corporeal schema or more appropriately here 'a coupling of body schemas' that incorporates the other person. Where the embodied adaptations necessary to be a competent guide became a habitual looking and feeling for others – an embodiment of their sight impairment.

It is interesting to note that even when not guiding these embodied adaptations – the 'coupling of the body schema' can persist. For example, I found myself navigating the landscape for two despite a conscious acknowledgement that the person I was guiding was no longer present. Such an experience indicates the way in which the brain is not simply a mirror like system but rather a device which becomes accustomed to certain patterns of sensory stimulation and attention.

This accustomisation, or habit of the brain, is evidence that below our intellectual control the mind is organising itself (Norretranders 1999, Connolly 2002). Here the guide experiences something comparable to having a 'phantom limb' where they still 'sense for two' and move as if they are still guiding another person. Thus it becomes important to acknowledge that habits can also guides people's behaviour. As Connely (2002, 36) puts it, 'brain regions respond not only to events in the world but also, proprioceptively, to cultural habits, skills, memory traces, and affects mixed into our muscles, skin, gut, and cruder brain regions'.

Here the idea of consciousness (in terms of an individual's continual conscious control of all their thought and actions) is a 'user illusion' (Norretranders 1999). Instead intuitions, habits and dispositions guide people's behaviours. Such observations undermine the idea that we have complete control and knowledge of our actions and intentions. Similarly, just as the guide experiences something comparable to having a 'phantom limb' partially sighted walkers incorporate sighted movements into their movements, habits and dispositions. For example, in order to navigate the steep and varied terrain of the Lake District and Peak District a silent relationship of mutual trust and proprioceptive, corporeal-kinetic understanding had to be established, which was reliant on touch. As Paul explains:

> When you hold somebody's arm, you sort of switch off and it's brilliant, you know? You can listen to the what is happening in the countryside, you can relax more... sometimes you are tense thinking you might get a cut in the head or you could fall over something that someone has put in the way. But really it is great the freedom really to walk unhindered with a guide. (Paul, blind since childhood, 35–45 years old; interview in June 2004.)

Here the synchronised movement of walker and guide that occurs through practice becomes a habit that enables Paul to 'switch off' and 'walk unhindered with a guide'; a coupling of the body-schema that emerges out of repetition and ultimately habits of practice. This practice of guiding and being guided involves a form of what feminist philosopher Diprose (2002) has referred to as 'corporeal generosity'. As she explains:

> Generosity... is an openness to others that not only precedes and establishes communal relations but constitutes the self as open to otherness. Primordially, generosity is not the expenditure of one's possessions but the dispossession of oneself, the being-given to others that undercuts any self-contained ego that undercuts self- possession. Moreover, generosity, so understood, happens at a prereflective level, at the level of corporeality and sensibility, and so eschews the calculation characteristic of an economy of exchange. Generosity is being given to others without deliberation in a field of inter-corporeality... (Diprose 2002, 4–5)

Guiding and being guided involves this form of generosity as 'being given to others without deliberation in a field of inter-corporeality'. This is a form of generosity and ethical sensibility which is embodied and which may remain at the level of habit. It is what Weis (1999, 131) has referred to as a 'bodily imperative' toward the good rather than stemming from a more precise set of ethical values and it involves an openness to transformation within a situation. It is comparable to the form of 'ethics as sensibility or ethos' that McCormack (2003) refers to in his reflections on the bodily ethics involved within the actions of dance movement therapy. In this therapeutic activity such ethics require 'an openness to the uncertain affective potentiality of the eventful encounter as that from which new ways of going on in the world might emerge.' (McCormack 2003, 503). This can also be seen as a form of bodily disposition that is open to the alterity of the other, where, as Weiss (1999, 163) suggests, we may be able to develop 'a sensitivity to the bodily imperatives that issue from different bodies as a necessary starting place for our moral practices'.

Yet despite having a grounding in the body, ethics as sensibility or habitual disposition is not always open to all bodies equally as an acknowledgeable possibility, for some bodies tend to be valued more highly than others. For example, I have shown that it is the sighted guides who must be 'recruited' and who symbolically occupy the 'worthy role' in these rural walking contexts. So like other forms of generosity, the *acknowledgement* of a bodily generosity or ethical sensibility remains governed by social norms and value. Such norms 'determine which bodies are devoid of property and so can only benefit from the generosity of others, and which bodies are worthy of gifts and which are not' (Diprose 2002, 9). The social norm that persists in this research context is that 'the visually impaired' should be the grateful recipients of the assistance from the 'generous and worthy' sighted volunteers. Yet I have shown that people with visual impairments also 'give' something of themselves in guide-walker relationship, including the 'gift of trust'.

Conclusion: The Touch of Two Bodies in Movement

In this chapter I have shown that the synchronous movements of walker and guide open up spaces of the body as a 'coupled entity' and opens a potential for a coupled, habitual 'ethical sensibility' to sense for two and move as one. This is an intercorporeal ethical sensibility distinct from the more standard social scientific notion of ethics as that which requires cognitive reflection or rule following behaviour (cf. McCormack 2003). Of course this co-emergent bodily sensibility is not unique to the walker-sighted guide relationship. Rather, it is a form of intercoporeal experience evident in a range of parenting and other caring roles. However, often such forms of corporeal experience have been undervalued, relegated to women's roles and ignored as a key source of ethical behaviour, therefore they remain deserving of attention here. Such observations are also significant for geographer's attempting to grapple with the non-representational qualities and affects of rural walking; for a renewed geography of disability that recognises the co-emergent nature of corporeal experience and for conceptions of ethical subjectivity as involving pre-reflective and habitual components. In these final paragraphs I summarise and reflect on these contributions.

Firstly, non-representational work on 'rural walking' has tended to start from a focus on the individual and their relations with the landscape (Lorimer and Lund 2003; Wylie 2005 and 2006) or on landscape and animal life (Lorimer 2006). This chapter builds on such non-representational work on walking and rural landscape, but develops a specific set of insights into the joint *corporeal attunements to each other* that are involved in walking through the material landscape together as walker and guide. Secondly, the chapter has shown how the *kinetic-synchrony* achieved by walker and guide involves a *coupling of their body-schemas*. Such an observation contributes to a renewed geography of disability which recognises visually impaired corporeality as not simply a way-finding problem or individual lived experience, but rather an intercorporeal experience which involves both (dis)abled and able bodies (Macpherson 2008; Power 2010; Crooks 2010). However, I think further work is needed here to explore the power relations that structure such joint corporeal engagements, including the gendered dimensions to corporeal generosity.

In this chapter the subject is found to act ethically 'in-the moment' via habit rather than reason. That is to say the original principles around 'doing good' that motivate guides to guide are not the ones that end up dictating all their bodily behaviour or enjoyment. Rather, they learn to guide and develop habits of guiding practice which result in a guiding sensibility as an embodiment of the other's sight impairment. Here guide and walkers bodies can be understood as simultaneously socially inscribed and capable of co-emergent, intercorporeal properties that produce alternative outcomes to that inscription. For people are not autonomous entities, instead their bodies are constantly being partly 'undone and remade' in particular circumstances and occasions (cf. Latour 2004). In the case of guiding and being guided it has been shown that new bodily habits can be adopted to suit the sightless group dynamic and new sensibilities developed within the guiding relationship that are attuned to the other's immediate needs. Geographers attempting to grapple with corporeality and ethical sensibility need to further develop their understandings of such co-emergent sensibilities.

Acknowledgements

With thanks to members of Sheffield Visually Impaired Walking Group and research participants from the *Vitalise* walking holiday groups.

References

Allen, C. 2004. Merleau-Ponty's phenomenology and the body-in-space encounters of visually impaired children. *Environment and Planning D: Society and Space*, 22, 719–35.

Allahyari, R. 2000. *Visions of Charity: Volunteer Workers and Moral Community.* Berkeley, CA: University of California Press.

Bissell, D. 2010. Passenger mobilities: affective atmospheres and the sociality of public transport. *Environment and Planning D: Society and Space,* 28(2), 270–89.

Butler, R. and Bowlby S. 1997. Bodies and spaces: an exploration of disabled people's experiences of public space. *Environment and Planning D: Society and Space,* 15, 411–33.

Caputo, J.D. 2003. Against principles: A sketch of an ethics without ethics, in *The Ethical,* edited by E. Wyshogrod and G.P. McKenny. Oxford: Blackwell, 169–80.

Charidi, E. 2009. How could blindness 'touch' the grounds of visuality?, Paper at *Visuality/Materiality: Reviewing Theory, Method and Practice Conference,* Royal Institute for British Architects, London, 9th-11th July.

Connolly, W.E. 2002. *Neuropolitics: Thinking, Culture, Speed.* London: University of Minnesota Press.

Cook, I. 1992. *Drowning in See-World? Critical Ethnographies of Blindness.* Unpublished MA Thesis, University of Kentucky.

Crang, M. 2003. Qualitative methods: touchy, feely, look-see? *Progress in Human Geography,* 274, 494–504.

Crooks, V.A. 2010. Women's changing experiences of the home and life inside it after becoming chronically ill, in *Towards Enabling Geographies: 'Disabled' Bodies and Minds in Society and Space,* edited by V. Chouinard et al. Aldershot, England: Ashgate.

Diprose, R. 2002. *Corporeal Generosity: On Giving With Nietzsche, Merleau-Ponty, and Levinas.* Albany, NY: SUNY Press.

Douglas, M. 1990. Foreword, in *The Gift,* by M. Maus. London: Routledge.

Golledge, R.G. 1992. Geography and the disabled: a survey with special reference to vision impaired and blind populations. *Transactions of the Institute of British Geographers,* 18, 63–85.

Harrison, P. 2000. Making sense: embodiment and the sensibilities of the everyday. *Environment and Planning D: Society and Space,* 18, 497–517.

Harrison, P. 2008. Corporeal remains: vulnerability, proximity, and living on after the end of the world. *Environment and Planning A,* 40(2), 423–45.

Hetherington, K. 2002. The unsightly: Touching the Parthenon Frieze. *Theory, Culture and Society,* 19(5/6), 187–205.

Hetherington, K. 2003. Spatial textures: place, touch and praesentia. *Environment and Planning A,* 35, 1933–44.

Hill, M. 1985. Bound to the environment: Towards a phenomenology of sightlessness, in *Dwelling, Place and Environment,* edited by D. Seamon and R. Mugaerauer. Dordrecht Press, 99–111.

Hull, J. 1990. *Touching the Rock: An Experience of Blindness.* London: SPCK Publishing.

Hull, J. 2001. Recognising another world. *Access, The National Journal for People with a Disability*, 3(2), 23–26.

Jacobson, R.D. and Kitchin, R.M. 1997. GIS and people with visual impairments or blindness: Exploring the potential for education, orientation and navigation. *Transactions in Geographic Information Systems*, 2(4), 315–32.

Keller, H. 1908. *The World I Live In*. New York: New York Review of Books.

Kitchin, R., Blades, M. and Golledge, R.G. 1997. Understanding spatial concepts at the geographic scale without the use of vision. *Progress in Human Geography*, 21(2), 225–42.

Kleege, G. 1998. *Sight Unseen*. London: Yale University Press.

Latour, B. 2004. How to talk about the body? the Normative dimension of science Studies. *Body & Society*, 10(2/3), 205–29.

Lorimer, H. and K. Lund K. 2003. Performing facts: finding a way over Scotland's mountains, in *Nature Performed: Environment, Culture and Performance*, edited by B.H. Szerszynski and W. Waterton. London: Blackwells, 130–45.

Lorimer, H. 2005. Cultural geography: the busyness of being 'more-than-representational'. *Progress in Human Geography*, 29(1), 83–94.

Lorimer, H. 2006. Herding memories of humans and animals. *Environment and Planning D: Society and Space*, 24(4), 497–518.

Macpherson, H.M. 2008. 'I don't know why they call it the Lake District they might as well call it the rock district!' The workings of humour and laughter in research with members of visually impaired walking groups. *Environment and Planning D: Society and Space*, 26(6), 1080–95.

Macpherson, H.M. 2009a. The inter-corporeal emergence of landscape: negotiating sight, blindness and ideas of landscape in the British Countryside. *Environment and Planning A*, 41(5), 1042–54.

Macpherson, H.M. 2009b. Articulating blind touch: thinking through the feet. *Senses and Society*, 4(2), 179–92.

McCormack, D.P. 2003. An event of geographical ethics in spaces of affect. *Transactions of the Institute of British Geographers*, 28(4), 488–507.

Merleau-Ponty, M. 1962. *Phenomenolgy of Perception*. London: Routledge.

Michalko, R. 1999. *The Two in One: Walking with Smokie, Walking with Blindness*. Philadelphia: Temple University Press.

Norretranders, T. 1999. *The User Illusion*. London: Penguin.

Olsezweski, B. 2008. El Cuerpo del Baile: The kinetic and social fundaments of Tango. *Body & Society*, 14(2), 63–81.

Paterson, M. 2009. Haptic geographies: ethnography, haptic knowledges and sensuous dispositions. *Progress in Human Geography*, 33(6), 766–88.

Power, A 2010. The Geographies of Interdependence in the Lives of People with Intellectual Disabilities, in *Towards Enabling Geographies: 'Disabled' Bodies and Minds in Society and Space*, edited by V. Chouinard et al. Aldershot, England: Ashgate.

Sennet, R. 2004. *Respect: The Formation of Character in an Age of Inequality*. London: Penguin.

Stoller, P. 1997. *Sensuous Scholarship*. University of Pennsylvania Press.

Swain J., Finkelstien, V., French, S. and Oliver, M. 2004. *Disabling Barriers, Enabling Environments*. London: Open University/Sage.

Thrift, N. 1997. The still point: Resistance, expressive embodiment and dance, in *Geographies of Resistance*, edited by S. Pile and M. Keith. London: Routledge, 124–51.

Thrift, N. 2007. *Non-representational Theory: Space, Politics, Affect*. London: Routledge.

Varela, F. 1999. *Ethical Know-how: Action, Wisdom and Cognition*. Stanford: Stanford University Press.

Vitalise 2007. *Sighted Holiday Makers*. <www.vitalise.org.uk/Visually-Impaired-Holidays/Sighted-Guiding.aspx>.

Watson N. 2003. Daily denials: The routinisation of oppression and resistance, in *Disability, Culture and Identity*, edited by S. Riddell and N. Watson. Harlow, England: Pearson.

Weiss, G. 1999. *Body Images: Embodiment as Intercorporeality*. London: Routledge.

Whatmore, S. 2003. Generating materials, in *Using Social Theory: Thinking Through Research*, edited by M. Pryke, G. Rose and S. Whatmore. London: Sage.

Wylie, J. 2005. A single day's walking: narrating self and landscape on the South West Coast Path. *Transactions of the Institute of British Geographers*, 30, 234–37.

Wylie, J. 2006. Depths and folds: on landscape and the gazing subject. *Environment and Planning D: Society and Space*, 24, 519–35.

Chapter 7

Touch, Skin Cultures and the Space of Medicine: The Birth of Biosubjective Care

Bernard Andrieu, Anne-Flore Laloë and Alexandre Klein

Introduction

Touch is at the core of therapeutic relations, yet the place of touch in medicine remains questioned. It is currently only thought of in terms of so-called alternative or 'soft' medicine. This chapter uses touch as an entry point for an analysis of touch's social influences on medicine because 'touch is our most social sense' (Field 2001, 19). Whereas touch therapies are becoming increasingly popular they have not yet found a secure 'place' in conventional medical practices, despite there being a social demand for touching in modern medicine; as such, the place of touch in medicine remains disputed and uncertain.

To address this, we first examine the place of touch in medicine, considering various possible paradigms to understand touch in relation to the body; we then argue for the need for a more holistic approach. As we show, by focussing on the place of touch in medicine it becomes possible to think of medicine as a tale of the skin. Therefore, by considering and integrating touch therapies, modern medicine can include the space of the body and the skin into medical ways of thinking. Certainly, as the border of the body, skin gives us a way of accessing the body's subjectivity whereas thinking about the body's space allows us to think about the physical space of the body and its relationship to care. Using this notion we seek to develop a theory of the Other in which the Other has a sensual and affective relation to the subject.

Moreover, we explore the ways in which alternative medicine is developing new conceptions of therapeutic touch. Originating in Europe and then exported to the United States in the early-twentieth century, alternative medicine subsequently returned to Europe with a new understanding of what touch is (Worthon 2004). We call this biosubjective touch, and use this notion to analyse the relationships between care and the body.

First, therefore, this chapter reviews the current relationship of touch to the epistemology of the body, focussing especially on how the body can be experienced through touch, and what this signifies. We then consider a number of strands of spatial touch which further examine some conceptual paradigms through which touch can be understood. This leads us to consider in particular the question of touch within the context of care of the body, or biosubjective care. In conclusion, we re-assess the present place of biosubjective care as a holistic way of experiencing the body and speculate about its future as a concept to make sense of place and space.

Touch and the Epistemology of the Body

The epistemology, history and philosophy of the body can be understood as the epistemology, philosophy and, most of all, history of a physical object. In this sense, the body is an interdisciplinary object whose modelling, like the problem of the relation of touch, is indeterminate as it occurs at the intersection of theoretical models that seek to fully codify it. As such, the body can be perceived as a moving, dynamic and lived object whose understanding has gradually been revealed across a range of human and social sciences.

Yet, ontologically and epistemologically, models of the body remain heterogeneous, focussing on different aspects of the material body, either physical or as interpreted culturally. The dispersion of models of the body in phenomenology, psychology or sociology in the beginning of twentieth century has produced an incomplete ontology of the body. Therefore, no synthesis of views can achieve anything more than describe either different characteristics of the body or the body in its biocultural expressions. Certainly, this fragmented epistemology produces a disunited ontology of the body. Whereas thinking of the body as a dispersed whole has been fruitful within contemporary human and social sciences despite an epistemological confusion and contradictory models which seek an exhaustive interpretation, the theory of emergence of touch, like a new object for the epistemology of the body and still used by many authors studying touch, has allowed us to think about the overall complexity of the body (Classen 2005). By acknowledging that neither touch nor the body can be fully reducible to scientific explanations, it has become possible to take advantage of the body's ontological fragmentation. However, the alibi provided by this epistemological disparity signifies that no genuine ontology of the body has been constructed, though the contributions of psycho-physiologists, psychosomatic doctors and medical practitioners in building theories of the physical practices of touch must be acknowledged.

Furthermore, the epistemology of the body can only be understood within the limits of deconstructivism which brings out the body's denaturalisation through the cultural determination of its practices. This means that the body's space, its cultural interpretations and its physical materiality must remain clearly differentiated when thinking about the body through touch. In particular, it is crucial to distinguish cause from effect, and think of the body in its different forms as distinct from how it is interpreted through touch. Certainly, the analysis of the body's interpretations through representations, beliefs and discourses reveals just how much both performance and recirculation are the actual driving forces behind the interpretation by the body of its situation. Nonetheless, it also emerges that historians of psychology have also significantly contributed to our understanding and cataloguing of these representations and behaviours.

The boundary between hermeneutic production of the body and interpretations produced by the body requires that the epistemology of the body systematically locates it in its physical act rather than understand it within a representation or already established discursive patterns. This means that it is important to differentiate between interpreting the body and interpreting what is sensed by the body in such a way that does not compromise the understanding of the body. For example, the risk of making the terms 'gender oppression' more ontological transforms the epistemology of the body into an ontology of the body by denying the gender characteristic of the interpretation of how the living is lived, or lived living (Butler 1991, 95). The body, the interpretation of the body and the body's interpretations are each considered; the body and its characteristics are therefore interpreted in themselves. Therefore, the consequences for touch (as a way of interpreting the different spaces of the body) in gender relations are a crystallisation of relational space such as the decomposition of different gender's places rather than the understanding of an intersubjective situation.

The advent of post-structuralism becomes problematic when the epistemological position questions the body and its interpretations. In order to deal with the dispersed body, as seen above, post-structuralism puts forward a distinction between levels of interpretation regarding what is primary, that is the body as an object, and what will always be secondary, that is its interpretative enactment whose representations are studied by the humanities and social sciences. Certainly, because of the limits of the body, its deconstruction remains within a strict culturalism that rejects its performative dimension in an historical situation;however, with touch we can suggest new interpretations of the lived or living body. The body can then be interpreted as a living object for philosophy, and, in this way, it is possible to base knowledge of it on the analysis of perception, feeling and emotion as perceived through touch. What we achieve is that the lived body and the living of the body are no longer divided and philosophies of the body can be interpreted through the unity between body and spirit. In this way, an emplaced body is no longer divided and can be read through this reconstructive philosophy that reconciles the otherwise dispersed body.

Following from this position, Duden (1987) argues that the study of conditions of emergence of the modern body cannot be separated from the study of the subject's manner with the feeling of his or her own body. She writes that though 'these two aspects of body history are inextricably intertwined, it is methodologically useful to separate them' (Duden 1987, 2). However, this difference between these two aspects cannot be methodologically separated because 'after all, my body determines my perceptions' (Duden 1987, 2). From our perspective, the problems associated with considering the body historically express themselves when considering the historical perceptions of women's physical senses. Indeed, as we discuss in a moment, the specificity of women's touch was included in the principle of creation of professional of care, thereby suggesting that gender should have an influence on the intensity, the orientation and the technique of touch. Following Duden's position that perceptions are undeniably affected by one's body, the position of femininity emerges as an important way of making sense of how the body can be interpreted and what its place is with regard to biosubjective care.

In this sense, the possibility of a physical perception of touch begins with the appropriation by the woman of medical knowledge and from her relationship to the doctor. As Duden (1987, 5) notes, 'the body is constituted only at the point where different fields of scholarship converge.' Thus, the body emerges from a history of progressive and successive discoveries, disallowing the idea of a single source of the object of the body or of spontaneous generation. This follows the Bachelardian difference between the historicity of the subject (here the experience of the body) and its formalisations; this differentiation implies that the historic reality of historians is a construction of the body, not a concept or a preconception of their constant nature. Indeed, Duden (1987, 7) writes that, '[i]n this way medical and scientific history becomes the story of how the 'real' body is 'discovered'.' From this point of view, cultural and social history can be said to examine how the body is discovered. Conversely, the history of psychology and psychohistory in turn place emphasis on the internalisation of influences. Thus, a part of women's history, along with some work from a patient-oriented history of medicine, can be read as 'the history of resistance to the well-intentioned, professional distortion of the body' (Duden 1987, 6). In this context, touch must therefore be described in the relational space between the patient and the health professional because the effects of touch are doubled and asymmetric within the situation of care. Crucially, this allows us to start thinking in new ways about skin as a new space where it functions as an essential site in relationships between patients and doctors. This provides us with a gateway to think about the body through using the skin as a new kind of space and build on Wegenstein's (2006) notions of bodily subjectivity.

A new space of skin

In his book, *Getting Under The Skin*, Wegenstein (2006) examines how the transformation of the body's subjectivity has been studied from three different perspectives: the deconstruction of the image of the body; the existence of a lived body; and the illusion of the modification of skin and the surface of subjectivity. While agreeing with this heterogeneity, we argue that the spaces of touch can be further analysed in conjunction with techno-science studies, the history of medical technologies and new media studies. Through engaging with new sciences, we highlight additional and significant relationships between skin and touch, and examine the perception the body in relation to these. At the centre of our concern is the relation of touch to the possibilities offered by technology in medical practices, especially cosmetic ones, and the evolution of the representation of the body's image in a media-driven society.

For the modern subject, the importance of the body's image is twofold. First, it is enacted by the constitution for the corporeal schema and body image; second, it hinges on the incorporation of body's norm in society. The study of touch in medical spaces highlights the intersection between the first and the second points because such practices as makeovers or cosmetic surgery and their subsequent care embody the possibility of the materialisation of an ideal body image through the hybridisation of natural matter with technical bio-design. The altered skin and the consequences of such medical practices are embodied not only within the space of the body but also in the conceptualisations of the body by the self. Therefore, the subjective transformation of skin by the alteration of biomatter produces new sensations that are physical, sensual and perceptual. The question we want to address then concerns how touch becomes a sensory experience by the modification of matter in media representations (e.g., cinematic, artistic, dramatic) and in the evolution of technology in terms of aesthetics, cosmetics and medicine (Taschen 2006). The strong link between the physical alteration of biomatter and consequences on the body to sensual perception then intersect within the body on a psychological level, as has been highlighted by new analyses of self in recent years.

Moreover, this link between the invention of cosmetic surgery and psychoanalysis is at the core of a new analysis of the correlations between surrealism, techniques of make-up and forms of absolute beauty in Hollywood cinema (Gilman 1999). Historically, the modification of male bodies after the First World War might be a good point for developing a new configuration of skin, since these consequences forcibly brought on a specific way of thinking about the body and its modified forms (e.g., Bourke 1996). More recently, the examination of the consequences of the digital turn on the virtual construction of an 'improved' bodily image is at the centre of a new conception of mechanism because mechanical suture finds, according to the Visible Human Project (VHP), a novel form of visible cartography of cosmetic possibilities for human being without the reference of tactile experience. By creating three-dimensional representations of so-called normal bodies, the VHP has created a way of visually imagining the body in detail that can aesthetically shape ideal notions of the body. These reconfigurations of the body, by showing the previously invisible interior of the three-dimensional body, presents itself as a possibility to appreciate new images of self since self-esteem will be better if the actual body is understood as being only one possibility of self that could potentially be altered at the level of stem cells or nano-cosmetics.

Returning to the question of gender and considering the analysis of images of the body in the representation of femininity, women and skin as undertaken in gender studies, the disarticulation of female bodies, especially in sexuality and performance, reveals how visual and narrative strategies to 'get under the skin' engage with new representations of bodily practices. This includes, for example, practices of self-mutilation, hostility towards pornography, and violence against the erotic and the aesthetic construction of femininity (Laurentis 1994). These techniques describe new representations of subjective skin and considering how the skin is felt subjectively. As Bourke (1996) notes, such representations are concurrent with the anticipated representation of what the art of the new modernity of the body showed during the course of the twentieth century. In fact, these ways of thinking about the subjective skin have been experimented with through bodily performances and the Abject Art Movement, including works by Carolee Schneeman, Genesis O. Orridge, GG Allin and Kira O'Reilly. These artistic representations can be compared for a characterisation of specific obscene cinema (Williams 2004) where thinking about the history of obscenity might be useful in defining the intensity of transgression under the skin as regards the delicate question of the female body's boundaries (Andrieu 2010) and the limits of the hybrid body. Through this lens, the female body, its relation to skin and how it is experienced and experimented with illuminates our understanding of how technology and cultural movements can affect and alter the manner in which skin is subjectively experienced both as an entity in itself and as a way of projecting one's self. This physicality of the skin emerges as a useful example of the place of skin in our culture and body. To complement this view, we will now consider the physical spaces of touch by considering the different ways in which the development of the brain as a physical object is useful within the context of biosubjective touch.

Brain, space and touch interaction

As we have shown both contextually and culturally, the human body must first be understood as interacting with its environment because it is itself receptive matter that is both informed and informing. As Heinämaa (1999) has asserted, the human body is not a monad but has a temporal character in phenomenology; if it were isolated from the world, it would be unable to construct its own self. Certainly, the body is only endowed with potential and has to meet the affordances of an environment to become real. In this sense, we might call it an interface, though the body does not remain passive. The body neither compliantly obeys the orders of the nervous system, nor is it solely an objective reflection of the world. Far from being a recording chamber, according to a mechanical metaphor, the human body is both a way in and out, through which the inside communicates with the outside and vice-versa. This is at the basis of the interactive model of brain development.

A model of interaction in brain development (rather than one of internal emergence) defines the brain's plasticity according to the cultural forms that determine the body itself (Andrieu 2006a). The brain undoubtedly has a singular connection to the human body within which it is situated. Whereas its natural elements are determined by genetic inheritance, the physicality of environmental surroundings presumably has to actualise and direct the plastic potentialities of the brain. Consequently, there is a specific relationship between the brain as a physical object situated within a body and the brain as a genetically determined entity. This gives the human body a central role in the materialisation of the brain, the effects of which will express themselves in neurobiological data. In this sense, nervous messages do not necessarily precede the esthesiological activity of the body: the brain cannot be the primary driver of the condition of the body and saying so would confuse cause and effect. As such, we can argue that there is no neurobiological basis to corporeal consciousness if we define corporeal consciousness as the relation of the body to the world, since it begins with intra-uterine life that, in turn, physically informs, directs and selects nervous networks and circuits. In this sense, if a human being is born with a brain which is not fully developed, and therefore whose role in defining consciousness is, also, unfinished, it can be said that this is so that culture can be the element necessary for its full qualification. The consequence of the brain's fragmented development is that the relationship between the physical and genetic brain is further shaped by the cultural locations of the later developments of the brain. Yet, developmental neurobiology fails to adequately link corporeal consciousness to the result of brain activity.

The body therefore emerges as a mediator for a number of reasons. The new models of a philosophy of the body, developed by a range of scholars in the United States (such as Hubert Dreyfus, Quassim Cassam, Mike Proudfoot, Samuel Todes) and in France (including Bernard Andrieu, Renaud Barbaras, Natalie Depraz, Michel Lefeuvre, Jean-Michel Roy, Francisco Varela) offer an interdisciplinary unification of the sciences to describe the relationship between the body, the brain and the mind within a physical materialism paradigm. Among others, the neuroscientists Alain Berthoz, Pierre Buser, Antonio R. Damasio, Marc Jeannerod, Joseph Ledoux, Vernon S . Ramachandran seek inspiration in the materialism of Denis Diderot and the monism of Spinoza: the action, the emotion, the representation and the decision are no longer naturalised and can, therefore, no longer be reduced to a biology of passions. This physical modelling of the brain has opened the possibility of neurocognitive phenomenology. What is physically lived by the subject is no longer separable from neurophysiological modifications of the brain and the nervous system. The relationship between the brain and the nervous system, then, is more than one of causality: the body informs the brain about sensations, communicating between the mind and the brain, and it expresses a relational affectivity with its socio-cultural environment. However, this position does not undermine the fact that the physiological or psycho-affective state of the body depends on hormones, the quantities of active neuronal networks and the quality of its genetic code. Certainly, the link between the unconscious and memory is also founded in the brain that records and reorganises representations in as many neuronal associations, while dreams, it can be said, remain the guardians of the individuation of brain. The historical involvement of a subject in the world is then more than the determinist result of, among others, endocrine hormones.

What this means is that we can then conceive of the human body as being a place of conversion, meaning that its form acts upon its matter, as we seen above with the brain. Similarly, its matter indefinitely determines its formation. This further signifies that we must distinguish between several subjective levels in order to understand, if not objectively explain, the modes of relations between the body and the brain, and touch is an important actor in this relationship. Whereas dualism allows for a substantial distinction between the body and the spirit without admitting to an intersection between the two dimensions, we posit that touch allows us consider these different aspects simultaneously and consider that each level corresponds to a level of representation for the human subject.

Considering the unconscious, we understand it as the subject's intimate representation that can be modified by the brain with psychoanalysis or following physiological damages. If we accept that a human being's brain enables them to represent themselves and develop representation by giving it a material support during the subject's activity, the effects of modifying the unconscious become entrenched within the brain itself. Furthermore, the interaction by neurophenomenology with the unconscious representation of the body in the brain, similarly to the neurological phenomenon of phantom limbs, offers a new possibility for a person to interact with a new representation of the physical body or touch his or her virtual limb (Ramachandran and Rogers-Ramachandran 1996). This parallel allows us to consider the place of the flesh which is, conversely, a dimension of affectivity upon which these feelings and imaginations feed into the subject's experience and the singularity of one's geohistory (Andrieu 2006b). Notions of unconscious, representations of the body and the body's physicality thus become entwined. We can then interpret what Merleau-Ponty calls the *corps propre* as the image of the body that results from the narcissist representation of the subject such as it sees itself or represents itself through others. This body is the objective body, the exterior body, the surface visible to others and the organism objectified by the mechanisms of techno-science. What this leads to is that we cannot say that the world is in front of the subject because the subject itself is inside the world. The boundaries of the body and the world are then blurred.

This overlap between the objective and subjective body itself 'subjectifies' in the sense that the matter of the body is the result of this constructive interaction. Through the term 'subjectifying' we are placing emphasis on the movement which makes singular each human body through its successive incorporations. Therefore, subjectivity is the result of adaptation and regulation through continuous movement, and the human body is subjectified from the moment of its conception through the interaction between the mother and child. Extending from this, we could say that nothing completely objective can be known, and thus knowledge of the world is relative to each human body. Yet, each person's relativity does not preclude scientists from approximately establishing a temporary 'truth' of the human body, since scientific truths that scientists are establishing cannot necessarily be removed from their own bodily perceptions.

Indeed, this knowledge must remain approximate because of the very constitution of the body: sensation is felt through the unique nervous structure of a body and therefore two human bodies can never feel the same intensity for the same object. Psychometry, the systematic measuring of sensation through experiment and modernised by electrophysiology, seeks to quantify the quality of experiences of what has been lived by the body. This method establishes an objective knowledge by measuring reaction times and perceptual thresholds and might be helpful in objectifying science. However, the visualisation of electric exchanges cannot reveal anything about the way the body feels them in itself. If the neurosciences have a good knowledge of how the body functions, what they demonstrate is the biological side of what is felt psycho-physiologically, or even exclusively psychologically, by the body itself, not the measured results of these sensations. The distinction between levels of description does not lead to a new dualism since one could not exist without the other. Nevertheless, should the body owe its activities to the strict application of neuronal networks, the contents of thought make the subject believe in a natural illusion of independence from the matter of its own body. Our point is not to deny the determining character of matter in body functions but rather to refuse the reduction of the organ to its function. In this sense, touch is unknowable and the complexities of the skin as a boundary and an organ of touch must be acknowledged all the while certain aspects of its expressions, such as electrical impulses, can be measured. Crucially, quantification must not be confused with knowing, especially as regards the place of skin as a vital organ of touch. Touch cannot be simply quantified, and doing so would mask the complexities of the psychological dimensions of touch, but quantification can be helpful as a way of understanding the complex relationships at play.

Returning to thinking about the materiality of both the brain and the skin, and through an esthesiology of the body proper, it follows that two human bodies can never feel the same intensity for the same object. This impossible intersubjectivity holds the bodies out of themselves and each body in itself, making it difficult to express linguistically what has been lived esthesiologically. The lack of objectivity is a component of any human body's relations with the world. However, any human body can at least be described from outside as an objective structure, but, following Andrieu (2007) we argue that developmental neurobiology must also describe a process of subjective individuation rather than only draw out a universal type of incorporation. Such a method would bring touch to the centre of the discussion and understandings of knowledge and experience through touch.

Touch as a somatechnics

With the concept of 'somatechnics', Andrieu's (1999) research on health and self described the effects of the incorporation of cultural and social techniques (such as the nationalisation of organ donor registers or blood donations) in the process of subjectification. According to Andrieu, touch as somatechnics brings a new experience of sensation through interaction with the world, since by practicing medicine on one's own body, the concept of medicine as a specific entity becomes blurred. The new possibility for the subject to turn back on itself the efficacy of cosmetic techniques (e.g., practices such as piercing, fitness, body-building) and in the depths of the skin (such as genetic innovation, cloning, experimentation with of stem-cells) is becoming commonplace. Following Mauss's (1934) concept of the 'technique du corps' to Andrieu's (1999) concept of somatechnics, there is a continuity in accounts of the effectiveness of technical incorporations of the body. The difference between Mauss' concept and Andrieu's is the decision of the subject to define its own body, even if the social norm is incompatible with this subjective choice.

According to Andrieu (1999), somatechnics are a set of instruments for the subjectification of the form and matter of the body. The body's agency is a new definition of self through the modification of the body: it is an imagined materiality that makes the individual body believe in the possibility of an intimate revolution through personal modification. This is due to the fact that the practices of personal modification or self-medication remain bound in the practices of consumer capitalism such as the pharmaceutical industry or advertising trends. Therefore, this agency remains an illusion from the point of view of capital organisation: bio-power proposes self-body control as a means to maintain panoptic domination through the body-work of the subject. The internalisation of bio-power by the patient transforms it into an agent of their own self health: just as the internal bio-power is exercised by each body-coach, performativity and performance are confounded in the same project, that is the auto-constitution of self.

However, for this to function, alteration of the self in order to achieve a norm cannot be systematically effected and somatechnics can be a practice of implicit normalisation accomplished by the subject itself. Considering these sorts of bodily alterations as a kind of physical deconstruction, which links into a poststructural sense of the term, of the body becomes helpful at this stage. For example, as regards discourses of deconstruction and gender, the confusion would be extended to the definition of touch. Indeed, the experience of touch implies an analysis of effect on the skin with the implication of a touch-agent such as intensity, violence or orientation. In this sense, touch itself can be a somatechnic for domination or for accompaniment. Further, the sense of care is different if individuals have a gendered consciousness of their role and power, meaning that touching can never be a neutral experience.

With the gendering and queering of the body, self does not consist of an essentialist definition which might be expressed by a somatechnics. Instead, the deconstruction of the body opens a new possibility of existence by the interaction of the body with a single situation. Indeed, Sullivan (2009, 314) argues that the use of the 'term somatechnics to think through the varied and complex ways in which bodily being is shaped not only by the surgeon's knife but also by the discourses that justify and conceptualise the use of such instruments'. The original contribution of Sullivan and Murray (2009) is the performativity of technique in the process of enfleshment of the self which centres on the touching of skin. The somatechnic is here a means to perform a potential self by a new body situation, as opposed to the conceptualisation of medicine of one's own body. However, this practice of enfleshment is not purely therapeutic, though the practices of therapeutic touch are also developed through enfleshment. We will now examine this concept of therapeutic touch and its place within medicine and biosubjective care.

The therapeutic touch

Therapeutic touch is a practice developed though techniques of nursing in U.S. maternity wards, according to Krieger (1975). Nursing consisted initially of techniques focused primarily on hygiene, which maintain and organise the environment of the patient. Florence Nightingale (1859, 39) describes this role as follows:

> Keep your patient's cup dry underneath. One very minute caution – take care not to spill into your patient's saucer, in other words, take care that the outside bottom rim of his cup shall be quite dry and clean, if, every time he lifts his cup to his lips, he has to carry the saucer with it, or else to drop the liquid upon, and to soil his sheet, or his bed gown, or pillow, or if he is sitting up, his dress, you have no idea what a difference this minute want of care on your part makes to his comfort and even to his willingness for food.

Nightingale sought to emphasise feminine knowledge of care within a rational, de-personalised and predominantly masculine institutional context. Following this view and though virtue, chastity and order and preservation, the nurse becomes part of the cure through adapted care:

> I have often been surprised at the thoughtlessness (resulting in cruelty, quite unintentionally) of friends or of doctors who will hold a long conversation just in the room or passage adjoining to the room of the patient, who is either every moment expecting them to come in, or who has just seen them, and knows they are talking about him. If he is an amiable patient, he will try to occupy his attention elsewhere and not to listen – and this makes matters worse – for the strain upon his attention and the effort he makes are so great that it is well if he is not worse for hours after. If it is a whispered conversation in the same room, then it is absolutely cruel; for it is impossible that the patient's attention should not be involuntarily strained to hear. (Nightingale 1859, 26)

A nurse's place is then defined by moderating their feelings and by showing patience; these qualities express themselves through a dexterity that is most often inherently attributed to women as if femininity was found in the use of good judgment within the intimate space of the cure. Nightingale's focus on the feminine illuminates understanding of touch by focusing it both on a tradition of professional care. The technically precise yet soft dexterity, attributed to feminine knowledge (though this view has prompted criticisms from feminists) nonetheless promotes a specific touch that is less feminine in essence than introducing into the hospital the recognition of tactile, non-invasive techniques of care. What we mean is that it is conceptually easier to attribute the origins of this touch to femininity (even if it is no longer gendered) than to including this specific way of caring to hospitals and their medicalised environment. These issues become evident and crucial to understand the place of therapeutic massage and touch.

Field (2001) has spent the last twenty years researching therapeutic massage. In 1992 she founded the *Touch Research Institute* at the University of Miami School of Medicine. In particular, she has examined the importance of establishing physical touch with newborn infants. Typically, stimulation of a newborn, especially those born prematurely, incorporates tactile-kinaesthetic, auditive, visual and social multimodal approaches (Field 1980, 302). Therapeutic massage (Field 1995a, 106) was experimented with during three sessions of fifteen minutes over a period of ten days with forty babies born at 31 weeks and weighing 1280 grams at birth. The massage was correlated with a weight gain in forty-seven percent of the cases. This provided convincing evidence of the effectiveness of touch in performance in comparison with Brazelton's scale. Field (1995b, viii) therefore concludes that touch in neonatal development facilitates interaction between the baby and the world, and provides a reliable grip to the exteriority of the world and self-respect.

That touch is inherently sensorimotor is assumed in the dynamic description of physical development. For example, support staff can help foster the relationship between a breast-feeding child and its mother by working to improve contact with the child, with the position of the mouth on the breast, through intentions and physical posture, and through an accompanying tactility. Touch is an integral aspect of this practice that tries to establish or restore a holistic relationship between a mother and her child through sensory contact. The physical care to the newborn provided by the mother distinguishes itself from the care of the neonatal team. This is because the care provided by the mother can be considered innate, whereas the medical team must consciously work to meet the demands for tactility and empathy. In other words, the medical team is in a position where it needs to acquire the ability to empathise with a baby and form a strong relationship with it while working on strengthening the relationship between the baby and its mother.

With this example, we highlighted the fact that to touch the body of others requires an ethics of care which considers the other, its own sensory intimacy and its emotional subjectivity. To take care of another requires techniques and an understanding of what is at stake in the space of the intersubjective communication of bodies. We can therefore talk of the therapeutic touch as deeply entrenched within space of care as embodied by bodies, practices and institutions. Empathy, understood as a manner of expressing this care, is practiced in the intersubjectives spaces of care, as seen for example, on maternity wards. This space of interaction between bodies and what each body requires in order for this relationship to function is crucial. Building on the examples discussed above, we suggest what we consider the minimal requirements of this care of touching both in the everyday environment and within the care space of the hospital.

The Care of the Body

Care of the individual is not reducible to a technical touch that carries out contact, movement and care where the body of the other is simply objectified. This sort of mechanical touch, or technical intention of touch (Illich 2005), would use the body like surfaces for technical applications without considering subjective implications, and we would not call that care. Indeed, the rigour of mechanical touch expresses itself through a coldness which lacks intersubjectivity and can be interpreted by a patient's body as an excess, if not an invasion, of body space.

Care is concerned with both *technê* and tact. In his *Anthropology* (1798), Kant defined logical tact to differentiate it from empirical tact, a prefiguration of the difference between professional and personal touch. Expanding from this notion, the reflection of logical tact represents the object on various sides and arrives at an exact result without being aware of acts which occur within its spirit. Therefore, professional touch uses logical tact, which is less emotionally empathetic than unconsciously psychological, to establish an immediate contact which will be translated into considerate care. This shift from spontaneous contact and considered care presupposes that understanding, more than affectivity, controls the body's intentions in professional touch in order to contain the emotion in the intention. The professional touch, conditioned by logical tact, is therefore a reflective touch that indicates the direction in which it thinks of the being in the world at the time of the effects of contact on the body of the others. What follows is that the notion of professional touch can extend into the one of personal touch through seeking to incorporate more personal qualities of touch within new models of professional care. This professional touch, though, bears the burden of being ascepticised in the situation of dealing with contagious diseases.

However, a system of contact accompanies the practices of emotional reassurance and this is the implied familiarity of nurses with patients. This system of contact is not possible insofar as each one gives up the right that one has to hold others remotely (Goffman 1967). Crucially, in this context, the right to touch is organised according to a hierarchy where any member of each group can touch others and be touched by others, but touching someone alters one's social category through the simple act of making one feel sympathy through a gesture. These asymmetrical relations between doctors and patients, doctors and nurses, and nurses and patients, alternate rites of presentation and rites of avoidance, are balanced by the provision of a touch of accompaniment and the creation of a compassionate bond. Because care applies to the human body, it affects not only the objectivity of the disease but also the subjectivity of the patient. To take care of somebody is thus not only a precaution of principle: the act is engendered very concretely through words and gestures, a communication between people. By looking after the body, the medical profession tests this communication intensely, as it is the link between the professional and the patient's body. The differentiation between professional and personal care here becomes blurred as practices of bodily care, relationships between patients, their bodies and the interactions with medical practices become entwined. We will now look at two ways in which the relationship between the patient's body and the medical profession expresses itself: body language and tactile perception.

Body language

Body language in medical consultations, even if accompanied by a conversation, distinguishes itself from extra-communicative gestures and is instead constituted of embodied gestures. On one hand, embodied gestures are quasi-linguistics because they are equivalent to the word but are acted. However, co-verbal gestures are more deictic and indicate a bodily segment to put the body in scene by the means of the pictographic and kineminic epic. Certainly, the question of correspondence or complementarity is important in considering body language. Significantly, the interpretation of body language found in a universal categorisation of postures and gestures helpfully provides access into emotional embodiment. For example, Ekman (1964) has studied emotional adaptation and argues that there are universal categories of emotions, associated to univocal facial expressions. He further defines joy, sadness, fear, anger, surprise and dislike as being the six basic emotions, and this categorisation is a useful tool to engage with how these core emotions might be interpreted within the expressed movements of body language.

For example, when thinking of the body's continuous accompaniment of life towards death, the connection between emotions and expressions of emotion shows how the body is sensitive to suffering and pain. Conversely, this identification between looking after the body and the *corps propre* happens at the same time of the process of being discharged from care and re-entering the world of the healthy and difficult separation between the patient and the person. In this sense, we can read in body posture the degree of emotion and we can develop, within the manner of caring, compassion or sympathy for helping the person in distress.

The incorporation of products, the imposition of gestures, the handling of tools, the affixing of artefacts certainly upsets the traditional co-ordinates of the representation of the body by the subject. One could test this instrumentation of one's own body through therapies of the body by experimenting with more treatment than is medically required. Conversely, one could also test this instrumentation of one's own body to the point of being able to represent that of others, even if, for looking after the body, all the difficulty is in the integration of this representation of the suffering of the other in its mode of action.

Significantly, looking after the body has the virtue of communication, both because of the immediate qualities of one's own body, but also because of the involvement and mode of action associated to it. Confronting the body of the other, and looking after that body, requires anticipating the feeling that it experiences so that a progressive reappropriation of the body is possible. This passage of the body from an *object of care* to the *body of care* places caring for the body in front of a singular responsibility, that is to transpose lives of the body looked after, reciprocally, in terms of the medical profession.

After all, behind and under the doctor's white coat, which can too easily be seen as a protective shield or a guarantor of statutory objectivity, exists a quiet, genuine body looking after another body and likely to incarnate within it its ethical acts as a supplement to the care of touch. Delicateness, attention, empathy and compassion can then be included into the body of care by including them into gestures, physical proximity and listening. These techniques of body division with others, even though they do not abolish differences and distances, make the moment and the space of the body of care essentially more sensitive. Including feelings of empathy and expressing them within the acts of looking after the body therefore restores emotive affectivity and the sensitivity to the practices of care.

A tactile perception

Through a predominantly physical care, everyone is touched in their psycho-physiology, if not their psychology: care of the body undermines the subject looked after by modifying the state of its body, perception of oneself, the imaged and diagrammed body, and modes of action. Yet, care of the body and restoring it to health is not necessarily the only goal, but the appropriation of oneself by the other also rests on a transmission of knowledge through gestures, techniques of the body and habits. Each gesture and posture becomes the expression of the singularity of a person and we can listen and help this person through the correct interpretation of their body language like the expression of their geohistorical situation. Indeed, care of the body is driven by a relationship in which one body alters another. However, this is despite, as we discussed above, only being able evaluate changes in nature, intensity and effects through the expression of physical, emotional and relational health. By entrusting one's care to someone other than oneself, this care can only be measured through tests which claim to define both the quality of the care of one's body and the state of the body.

The development of tactile perception (Schiff and Heller 1991, Schiff and Foulke 2004) is enacted through interactive observations of nonverbal behaviour. In geriatric nursing, for example, according to Saulnier (1989, 37), such observations include,

> [m]ovements of the head: distance, shaking, bringing together, tremor…; facial mimicry: grimace, crumpling of the face, smile, tics; movements of the eyes: seek the glance, flow of tears, closing of the eyes, fixing of an object or a person, escape of the glance, tears to the eyes, opening of the eyelids; gestures of the hands: absence of movements, agitation of the hands, crossing of the hands, gives the hand, gripping of an object or hand, sprouting of the finger, research of the hand of looking after, withdrawal of the hand of looking after, handshake of looking after, tighten the hand, tremors, etc.

The observation of one or some of these signs enables nurses to assess patients mostly through observation.

In addition, within the specific context of AIDS, the direction of contact in looking after is also confronted to a physical distance and a fear of contamination, therefore requiring a need for an indirect tactile touch. This asceptic touch, care and the notion of 'looking after' still maintain distance and non-tactile contact. Here there is then is confusion between physical touch and possibility of contamination, understood like a contagion. Yet, removing the plastic glove preserves a contact other than the properly ascepticised one by taking responsibility for the risk of a real meeting with either the other person or with the patient. Therefore, the visual dehumanisation of the instrumentalised objectivation of the body by the technicality of the gesture can be attenuated by a phenomenology of the touch in the medium of care (Vinit 1999). Certainly, to touch the body of the patient directly would restore a true communication as if the skin, little by little, could, through the true risk of contamination, restore authenticity and truth.

Conclusion

This chapter has sought to bring together a number of ways of interpreting touch in medicine around the notion of biosubjective care. By examining touch in relation to the epistemology of the body and tying different approaches to discourses of care of the body, it focused on a number of issues that are at the core of the tensions between medicine, touch and the self. For example, physical space is involved with touch from the first stages of the development of the body while the importance of acts made by the body remains deeply relational. At the same time, the role of perception implies a new conception of medical care through the interconnection between physical space and personal relations. By this model, the body can touch a person in care by the mediation of professional situation, though this act of touching is not, as we have shown, straightforward. Furthermore, the spatiality of touch is an experience of the self and for relationships to others, shaping physical and emotional relationships, as well as personal and professional ones. If we are at a sufficient distance to touch the other body, we can interpret gestures, posture and body language: these attentions and care for skin and the surface of the body find all the somatechnics incorporated under appearance. What this leads to is that within spaces of medical consultations, the diagnosis now has a visual and virtual dimension.

Together, what these considerations lead to is that touch embodies a medical model of the body that reaches it according to a topology which is not summarised solely with the sounding of a stethoscope. It becomes crucial to transcend the paradigm of a visual palpation whose semiology requires only one sensory triangulation (sight-touch-hearing), thereby reinforcing 'the glance bordering on the touch' (Foucault 1967, 166).

At the same time, touch remains a therapeutic mode that belongs to us so that we can enact one of the biosubjective activities of the body. This is what we considered through the notion of somatechnics, stressing the importance of being the doctor of one's own body. By discussing being doctor of one's own body, as opposed to being a doctor to oneself, we described the dynamic models by which the subject uses the processes of its body to improve its health, and its shape, if not also the matter even of its body. This highlighted tensions between self and care while acknowledging the importance of considering one's self as well as one's care.

Biosubjective care emerges, therefore, as a multifaceted discipline since it involves the physical and the emotional, self and other, personal and professional, immutable and moveable, all of which need to be taken into account in the delivery or self-delivery of care. This further implies that the quality of touch matters both for the reconfiguration of the surface of one's skin and for the modification of the body space: the effects on perception and sensation are embodied from the brain to the expression in language through empathy, care and compassion. Through a full understanding of touch and the place of touch, touch provides security and confidence by the incorporation and the recalibration of our self-help. With touch, the spatiality of medicine is expressed internally as well as interacting with the degree and the intensity of physical relation. This is at the core of biosubjective care and what we propose as a way forward to understanding and incorporating care into medicine.

References

Andrieu, B. 1999. *Médecin de son Corps.* Paris: PUF.

Andrieu, B. 2006a. *Le Dictionnaire du Corps en Sciences Humaines et Sociales.* Paris: Editions du CNRS.

Andrieu, B. 2006b. Brains in the flesh. Prospects for a neurophenomenology. *Janus Head: Journal of Interdisciplinary Studies in Literature, Continental Philosophy, Phenomenology, Psychology and Arts*, 9, 129–49.

Andrieu, B. 2007. Embodying the chimera: biotechnology and subjectivity, in *Signs of Life, Bio Art and Beyond*, edited by E. Kac. Cambridge, MA: MIT Press, 57–68.

Andrieu, B. 2010. Toward pluripotential hybrid skin. *Shifter*, 16, 37–44.

Bourke, J. 1996. *Dismembering the Male: Men's Bodies, Britain and the Great War.* Chicago: University of Chicago Press.

Butler, J. 1991. Disorderly woman. *Transition*, 53, 86–95.

Classen, C. 2005 *The Book of Touch*. Oxford, Berg.

Duden, B. 1987. *The Woman Beneath the Skin: A Doctor's Patient in Eighteenth-Century Germany*. Cambridge, MA: Harvard University Press.

Ekman, P. 1964. Body position, facial expression and verbal behavior during interviews. *Journal of Abnormal and Social Psychology*, 68(3), 295–301.

Field, T. 1980. Supplemental stimulation of preterm neonates. *Early Human Development*, 4(3), 301–14.

Field, T. 1995a. Massage therapy for infants and children. *Developmental and Behavioral Pediatry*, 16(2), 105–11.

Field, T. 1995b. *Touch in Early Development*. Mahwah, NJ: Lawrence Erlbaum Associates.

Field, T. 2001. *Touch*. Cambridge, MA: MIT Press.

Foucault, M. 1967. *Birth of Clinic: An Archaeology of Medical Perception*. Paris: PUF.

Gilman, S.L. 1999. *Making the Body Beautiful: A Cultural History of Aesthetic Surgery*. Princeton, NJ: Princeton University Press.

Goffman, E. 1967. *Interaction Ritual: Essays on Face-to-Face Behavior*. New York: Doubleday.

Heinämaa, S. 1999. Simone de Beauvoir's Phenomenology of Sexual Difference. *Hypatia*, 14(4), 114–132.

Illich, I. 2005. *La Perte des Sens*. Paris: Fayard.

Kant, I. 1798[2006]. *Anthropology from a Pragmatic Point of View*. Cambridge: Cambridge Texts in the History of Philosophy.

Krieger, D. 1975. Therapeutic touch: the imprimatur of nursing. *Amererican Journal of Nursing,* 75(5), 784–87.

de Laurentis T. 1994. *The Practice of Love: Lesbian Sexuality and Perverse Desire*. Bloomington, IN: Indiana University Press.

Mauss, M. 1934[2010]. Techniques du corps, in *Philosophie du Corps*, edited by B. Andrieu. Paris: Vrin.

Nightingale, F. 1859. *Notes on Nursing: What it is and what it is not.* Philadelphia, PA: J.B. Lippincott.

Ramachandran, V.S. and Rogers-Ramachandran, D.C. 1996. Synaesthesia in phantom limbs induced with mirrors. *Proceedings of the Royal Society of London,* 263(1369), 377–86.

Saulnier, D. 1989. Les cris répétitifs. *L'infirmière Canadienne*, 85(11), 35–38.

Schiff, W. and Foulke E. 2004. *Tactual Perception*. Cambridge, Cambridge University Press.

Schiff, W. and Heller M.A. 1991. *The Psychology of Touch*. Mahwah, NJ: Lawrence Erlbaum Assocociates.

Sullivan, N. 2009. The somatechnic of intersexuality. *CLQ, A Journal of Lesbian and Gay Studies*, 15(2), 313–27.

Sullivan, N. and Murray, S. 2009. *Somatechnics: Queering the Technologisation of Bodies*. London: Ashgate.

Taschen, A. 2006. *La Chirurgie Esthétique*. Paris: Taschen.

Vinit, F. 1999. Isolement et contact dans la maladie: esquisse phénoménologique autour du sida, in *Corps et Science: Enjeux Culturels et Philosophiques*, edited by L.P. Bordeleau and S. Charles. Montréal: Liber, 133–146.

Wegenstein, B. 2006. *Getting Under The Skin: The Body and Media Theory.* Cambridge, MA: MIT Press.

Williams, L. 2004. *Porn Studies*. Chicago: University of Chicago Press.

Worthon, J.C. 2004. *Nature Cures: The History of Alternative Medicine in America.* London: Oxford University Press.

Chapter 8

Touching Environmentalisms: The Place of Touch in the Fraught Biogeographies of Elephant Captivity

Jamie Lorimer

Thilak: My name is Thilak and I am a mahout. I am 42 years old and this is my elephant, Somalia. I inherited her from my father and came here with her to the sanctuary when she got weak. She is an easy animal and does not usually cause any trouble. She is old and is good with the tourists. My father taught me how to speak to her and she obeys me when I touch her. I know when she is tired or ill and when she is hungry I feed her. I take her to the water and let the tourists wash her. This is my job, it is not good but it is all I can do.

Sally: My name is Sally and I love it here. I can hug Somalia all the time when I am washing her. I am really lucky because I get to sit on her. She won't lie down so I get on top of her and I walk down to the water, and then she does her spray, and I can lean forward under her body, and lie on top of her, and I can scrub her, and I can kiss her, and I can whisper in her ear. I love that. I love that. I get right on top of her and I can kiss her and tell her that she is beautiful and that I love her and the pain will go away. It makes me cry and I get sick of scrubbing because my arm kills but being close to her, I love that. I don't think Thilak understands, but it is so spiritual.

Somalia: They call me Somalia. I am an elephant. I was captured from the forest when I was young and taught to fear and obey men with sticks. I have lived with people most of my life. I used to carry logs but now there is no work and I am chained to this tree by the river. Many people come and stare and talk at me. Sometimes I carry them on my back and they feed and wash me. I try and follow orders from the mahouts, but my sight and hearing are going. I am bored. I am lonely and I am afraid. I rock myself to take the pain away. I am 60 years old and will not live for much longer.

Sally, Thilak and Somalia are three protagonists whose lives and bodies touch and become enfolded in the multi-species and multicultural ecologies of an Asian elephant sanctuary in Sri Lanka.[1] Sally is an Australian tourist who has travelled to stay for several months. Touched at a distance by elephants on film, she pays for touching encounters designed to rehabilitate Somalia from a life of hard work. Her donations help sustain the captive elephants and mahouts that live at the sanctuary. For Thilak, touching and being in touch with Somalia is imperative for his corporeal and economic survival. Interspecies communication with a large and dangerous animal depends on his attunement to Somalia's mood and his judicious touch of familiar body parts. Somalia is a 'companion species' (Haraway 2008), habituated to interspecies living. With poor sight and hearing she depends on touch and smell for communication, stimulation and sustenance. Touch – in different forms – is central to this triangle of relations.

This triangle of touching encounters provides a compelling example of an increasingly popular and powerful mode of environmentalism, whose affective economies centre on touching encounters between affluent consumers and captivating animals. The promise of haptic encounters with wild and captive animals has stimulated significant growth in forms of domestic and international nature-based tourism (Bulbeck 2005, Curtin 2005). Historically popular forms of visual entertainment associated with safari and zoos (Davis 1997, Mullan and Marvin 1987) have been supplemented by opportunities to swim with dolphins, pet turtles, hold monkeys, and ride and wash elephants (Cloke and Perkins 2005, Gray and Campbell 2007, Kontogeorgopoulos 2009). While, growing numbers of people like Sally, travel and pay to work hands-on researching and rehabilitating animals (Cousins et al 2009, Lorimer 2009). In identifying these touching environmentalisms I am not claiming the emergence of entirely new ways of engaging with animals. Bodies, with their diverse materialities and repertoires of skill, sense and affect have always played a key role in the generation and contestation of natural knowledge; what has changed is the ways in which important groups of people think about the role of touch and legitimate and contest their claims to knowledge.

This trend is significant in the context of the changing political economies of contemporary environmentalism, which have seen a decline in national and multilateral government spending on conservation, matched and sometimes exceeded by a growth in the power and resources of non-governmental organisations and private companies (Rodriguez et al 2007). These organisations are directly dependent on the mediated affections of affluent (frequently Western) citizen-consumers for their resource base. This trajectory is powerfully illustrated in the pre-eminence given to ecotourism in the emerging discourses and practices of free-market wildlife conservation (Brockington et al 2008, Duffy 2002, Honey 2008). Here science and entertainment meet as funds flow from high profile advertising campaigns and commodified encounters with diverse natures (Brockington 2009).

---

1   A short film that I made that visualises one of their encounters can be seen at <www.youtube.com/watch?v=SQBoZzLVtbA>.

At best, captive animals act as sacrificial flagship species, drawing in resources for their free-ranging and/or less charismatic kin and local human co-inhabitants, while offering didactic platforms for cosmopolitan environmental education. At worst, these animals lead short and impoverished lives in conditions far removed from those for which they have evolved. Paradoxically, one of the consequences of the rise of touching environmentalism has been the proliferation of captive animal attractions and commodified opportunities for proximal encounters in the 'wild'. Furthermore, the affective international political economies that these institutions help support could perpetuate exploitative political ecologies associated with colonial conservation (Adams and Mulligan 2003).

In her recent book *When Species Meet* Donna Haraway (2008) starts with touch to open up a series of timely questions about the fraught 'past-presents' and 'responsibilities' that haunt contemporary encounters between humans and animals like these. She argues that we cannot understand touch in bounded isolation but must attend to the distant but connected spaces and times through which it is framed. In this chapter I take the encounters between Sally, Thilak and Somalia as a one compelling example of this mode of touching environmentalism, which has long and complex historical geographies. I draw on Haraway and a growing body of work on touch at the intersections of animal studies, cultural geography, film theory and biophilosophy (Colebrook 2005, Derrida 2005, Marks 2002, Paterson 2007). I first map the modalities of touch associated with the triangle of human-elephant companionship at the elephant sanctuary in Sri Lanka. I then contextualise these modalities to account for the rise of touching environmentalisms before critically reflecting on the politics and ethics of touch in the context of debates over the management of captive elephants. In conclusion I briefly outline some of the implications of attending to touch for environmental thought and practice and appeal for further research.

Modalities of Touching Elephants

Recent work at the intersections of animal studies (Bekoff 2002), biophilosophy (Ansell-Pearson 1999), science studies (Haraway 2008) and cultural geography (Whatmore 2006) has begun to develop 'more-than-human' approaches for rethinking environmentalism. Touching bodies are central to a shared relational ontology that is open to the attachments, crossings and ethics (Bennett 2001) that pass between humans and nonhumans. Conjoining techniques from ethology and ethnography (Lestel et al 2006) these studies have developed and applied innovative methodologies for witnessing human-animal interactions that go beyond prior preoccupations with vision. Animal encounters are understood to comprise skilled processes of 'learning to be affected' (Despret 2004) that draw on: the full range of senses – including touch, movement and balance; non-verbal means of interspecies communication – including comportment, expression, gesture and mood; and the affective energetics of these encounters actualised into emotions like love, awe, fear and empathy. Existing studies have explored modes of human-animal companionship including dog-walking (Goode 2007, Laurier et al 2006) and racing (Haraway 2008), bird-watching (Lorimer 2008), horse-riding (Game 2001), wildlife conservation (Hinchliffe et al 2005) and animal-herding (Lorimer 2006).

I employed these methods for disentangling some of the modalities of touch performed by Sally, Thilak and Somalia in ethnographic fieldwork carried out in Sri Lanka in 2006-7. Here I filmed, observed, interviewed and participated alongside approximately thirty volunteers, who worked with eight elephants and their mahouts at an elephant sanctuary. The volunteers were mostly female (46 out of 55 between 2005–7), British and aged between 18–40. On average they spent at least four weeks at the sanctuary, washing and feeding 'their' elephant. The mahouts were all men and came from mahout families, preceded and taught by their fathers. Each looked after a single elephant which had either been inherited from a previous generation or acquired at a young age. They lived in close proximity to their animal but rarely own it. As opportunities for physical labour have declined, elephants are now expensive status symbols for a wealthy elite. The mahouts are poorly educated, lowly paid and their work is dangerous; life expectancy is short. The elephants at the sanctuary are mostly old, retired animals. They often have health problems and are kept chained to confine them to the sanctuary and to ensure visitor safety. They are released when they are being washed or are carrying visiting tourists on elephant rides.

The first modality of touch in this knot of interspecies relations concerns the affect of Somalia – and of elephants more generally – on people. Elephants are extremely popular. They are the most researched of all wild animals (May 1988) and feature prominently in Western and Sri Lankan culture, commerce and conservation (Jayewardene 1994b, Scigliano 2004, Wemmer and Christen 2008). This ability of elephants to touch and move diverse groups of people stems in part from their 'nonhuman charisma' (Lorimer 2007). This is a multi-facetted and relational property that encompasses the anatomical, ethological and aesthetic properties of the animal, as well as the affective energetics generated within interspecies encounters. Asian elephants like Somalia are charismatic species par excellence. They are large, visible and easily differentiated organisms. Anatomically they are both familiar and peculiar – they have four legs, two eyes and a tail, like most domestic pets and livestock, yet they also have a trunk and disproportionately large ears. Ethologically elephants are distinguished by their intelligence – or sagacity. These are animals with societies and multiple cultures; they have complex emotions and can adapt to living in close contact with humans (Sukumar 2003); they provide a source of fascination.

Popular evocations of elephants tend to focus on their eyes and trunk (Rothfels 2008) in stories centred on family dynamics (Lorimer 2010a). This framing serves to anthropomorphise the animal, highlighting familiar anatomical features that resonate with the corporeal triggers identified by philosophers like Levinas and Heidegger for the differentiation and elevation of the ideal ethical human. Handy

Heideggerian (Baker 2003) elephants, equipped with dextrous trunks can use tools and feel for others. Elephants with binocular vision, express faciality and offer a reciprocal Levinasian gaze (Jones 2000). As familiar beings, animated with such features these animals can easily be assimilated into powerful (humanist) frameworks for ethical concern (Lorimer 2010a).

For many of the Western volunteers, the charisma of the Asian elephants at the sanctuary had been framed in advance by powerful evocations on moving imagery. They had been touched at a distance by the 'fingery-eyes' (Hayward 2005) of wildlife film, which drew them to the species and established 'anticipatory knowledges and expectations' (Cloke and Perkins 2005, 910) and performative repertoires for human-elephant interaction. For example, Amy – a British volunteer in her early 50s – explained how she was inspired to work with elephants by the BBC television series *Elephant Diaries*, in which the wildlife celebrity Michaela Strachan narrates stories of the rehabilitation of individual orphaned African elephants and their return to the wild. The affective logic of this television series evokes elephants as tragic, doleful and suffering victims restored to life by the touch of patient, loving humans. Emulating this logic Amy was fixated by her elephant's eyes and hypnotised by its swaying trunk. She would sit for hours talking and staring at Rani, her chained animal, positioned just out of trunk's reach. She explained to me that:

> I love to look into Rani's eyes and she listens to me. You can just stand and I talk to her and she loves it. She rumbles away. I mean how many people get to do that to talk person to person to an elephant? I like to try and counteract the shouting and the sticking and the hard life she's had to give her a nice little bit of soft talking and gentle lulling, if you like. It makes me feel calm.

Although she had never owned or worked with animals, Amy had been drawn to the elephant sanctuary for a respite from an emotionally difficult life back home. She was seeking a therapeutic holiday amongst like-minded people and in close proximity to gentle and threatened animals. She felt an affinity with Rani and expressed a desire for mutual salvation.

Amy's mediated vision of elephants is situated within a wider sensory matrix, which has been the subject of much recent writing by film theorists interested in touch (Barker 2009, Marks 2002, Sobchack 2004). Repeated viewings and responses to the BBC's moving televisual imagery had framed Amy's 'optical unconscious' (Doel and Clarke 2007) toward elephants and had viscerally enfolded her body within the material world through a performative and affective process. Amy had learnt to be affected by elephants through her televised encounters with Michaela Strachan, which resulted in an unsystematic, confused but nonetheless power disposition towards Rani and the other elephants. Here we see the 'haptic aesthetics' (Colebrook 2005, Paterson 2007) and diverse 'affective logics' (Carter and McCormack 2006) that animate touching evocations of elephants on moving imagery. These evocations catalyse the charisma of the animal on film, triggering contagious affections and actions that can have important political and economic consequences (Lorimer 2010a).

In contrast, for Sally the charisma of the elephant and its recuperative potential lay in corporeal immersion. Sally explained that as a skilled and experienced horsewoman her encounters with Somalia were initially orientated by her familiarities with touching and responding to horses (Game 2001). She was disconcerted by the horizontal oscillation of her walking elephant, which differed from the vertical rhythms of a trotting hose. Similarly, the topography of sensual places for equine petting did not map onto Somalia's unfamiliar facial anatomy. She strove to clean and scratch behind the ears, or to stroke the side of the head – as she would with her horse – but she was unsure of the response this generated. She found it difficult to gauge Somalia's mood and to interpret the demonstrative movements of the alien trunk. Her strange vocalisations opened a communication gap, triggering curiosities that could only be partially satisfied by Thilak's hesitant English. She explained that she was only just beginning to attune to Somalia and learn to be affected by her behaviours. However, touch remained the pre-eminent sense in her encounters, where the daily highlight was the elephant bath. Lying semi-naked astride Somalia in the river, Sally scrubbed while the elephant sprayed water over her back. Afterwards, she explained how with her body conjoined in these rhythmic caressing, scrubbing and drenchings she entered into a trance – becoming-centaur in a haptic moment of recuperative reterritorialisation (Fullagar 2000).

In contrast to the experimental, uncertain and therapeutic modalities of touch performed by Amy and Sally, Thilak touched and was touched by Somalia with the time-deepened familiarity of an experienced mahout in the company of a working animal. He explained that his training had begun as a boy, when he would watch and help his father working with elephants in the forest. He had learnt the vocabulary that is used in Sri Lanka to communicate with trained elephants, which comprises over a hundred oral commands and a topography of pressure points on the animal's body that can be touched with a foot or the *ankus* – a stick with a metal point and hook (Hart and Locke 2007, Lair 1997). However, each elephant is unique and has complex moods and idiosyncratic behaviours. Thilak explained how he learnt to attune to Somalia over time, sensing and responding to her oral, olfactory and haptic cues. Her grunts, gestures and secretions constitute an interspecies language through which communication takes place and their relationship is negotiated. Thilak knows when Somalia is angry, hot, hungry, happy and sad – indeed his life depends on it. He is wary about dealing with unfamiliar elephants, especially any with a history of violence. He respects Somalia as a working animal; their relationship is characterised by discipline, respect and a degree of empathy.

From the little that is known about Somalia's biography it appears that she was captured in the East of the country in one of the last elephant 'khedahs' that accompanied the post-war development of the Mahaweli catchment (Jayewardene 1994a). During a khedah free-ranging elephants were driven into a stockade (Katugaha 2008) and young animals are then removed from their mothers to be 'broken' with the use of drugs, ropes and protracted exposure to noise, lights and dousing (Sanderson 1896, Williams 1950). A team of trainers will crowd the animal, smothering its body with their own. This habituates them to human contact and establishes dominance; inserting the mahout as a social superior in the elephant social order. Broken elephants are then chained and manually manoeuvred into desired positions through the painful application of an ankus. Through this 'free contact' method of elephant conditioning and management Somalia learnt to connect body positions and behaviours to commands given when the ankus is applied. Once obedience is secured then the ankus is no longer applied, but the threat remains; Thilak was rarely to be seen without his ankus.

An elephant like Somalia that has been raised in captivity becomes a different animal to one free-ranging in a national park (Whatmore and Thorne 2000). Recent research by ethologists suggests that captive elephants kept under a free contact modality of touch live in a state of 'learned helplessness' – physically able to resist human instruction but too terrified of the consequences to risk disobedience (Clubb and Mason 2003). Somalia is an intelligent, emotional animal who is aware of herself and her history (Sukumar 2003, Varner 2008). She possesses highly developed senses of touch, smell and hearing and lives in haptic, olfactory and aural environments alien to most human sensibilities. Somalia's natural curiosity is tactile and gustatory and is expressed through trunk and mouth explorations. She has spent much of her recent life chained at a standstill with diminished stimuli and contact with her kin. Elephants have never been domesticated (Clutton-Brock 1999); in contrast to several other large herbivores, they do not breed in captivity. Like many free-ranging, sociable and intelligent animals they are badly affected by solitude and confinement. Thilak explained that Somalia – and other elephants like Rani – had grown frustrated and were starting to develop stereotypic behaviours; the rhythmic weaving that so hypnotised Amy is a coping response to chronic boredom (Elzanowski and Sergiel 2006).

Touching Historical Geographies

As Donna Haraway (2008) has made clear, to contextualise these encounters – and to understand the shifting place of touch in (post)colonial environmentalism – we must attend to the distant but connected spaces and times through which they are framed and by which they are haunted. There is a growing interdisciplinary interest in more-than-human geography, anthropology and archaeology in attending to the historical traces that are animated in the bodies, gestures and emotions performed in interspecies encounters. Diverse work on landscape (Ingold 2000, Lorimer 2006, Yusoff 2007), assemblages (Legg 2009, Lorimer and Whatmore 2009) and materialities (DeSilvey 2007, McCormack 2008, Tilley 2004) provide useful methodologies for attuning to the historical geographies of response-ability (Haraway 2008) through which the past touches the present.

The touching encounters between Western visitors, mahouts and Sri Lankan elephants have long and diverse spatial histories too numerous to fully report here. In one compelling example, elephants feature prominently in the accounts of early British visitors to the island, who from the seventeenth century onwards recount fabulous tales of the animal's anatomy, behaviour and importance to local cultures (Lahiri-Choudhury 1999). For many of these (largely male and colonial) authors the elephant was the king of beasts, the noblest of the game animals and most worthy of skilled bodily pursuit, fearsome combat and heroic death (Lorimer and Whatmore 2009). For nineteenth century advocates like Samuel Baker (1854, xiii), elephant hunting required a 'fine and deft touch'; a detailed and hard-earned embodied sensibility for the ecology and behaviour of one's prey. The sportsman – who embodied all of the virtues of Victorian masculinity – was touched by his animal opponent and strove to ensure their swift, dignified and painless demise (Animal Studies Group 2006, Mackenzie 1997).

This touching sensibility of hunting was much derided by a divergent strand of elephant naturalism that became ascendant as the nineteenth century wore on. For James Emerson Tennant – colonial natural historian and proto-scientist – the corporeal proximity and viscerality of elephant hunting in Ceylon rendered it both unmanly and unwholesome (Tennent 1867). Men like Samuel Baker, caught with blood on their hands and with their passions running high, transgressed the stern identity of the Victorian colonial sahib (Collingham 2001) and the disembodied ideal of an enlightened man of vision (Pratt 1992). Such hunters were too close to lively bodies for their own good. For Tennant and other natural historians elephants (preferably dead ones) should be dissected and revealed, or observed at a distance through scoping technologies, maps and diagrams. Tennant and his contemporaries relegated touch and elevated vision as the route to natural knowledge (Haraway 1989).

Meanwhile, when they returned from the tropics to the salons of the metropolis both Baker and Tennant would have witnessed the growing concerns for animal welfare amongst the urban middle and upper classes. With Queen Victoria as its patron the Royal Society for the Prevention of Cruelty to Animals (RSPCA) was helping to catalyse moral outrage at animal cruelty amongst its powerful members and encouraging enthusiasms for social reform (Turner 1980). Their interventions focused on the treatment of domesticated animals by the colonised and working-classes and were characterised by disdainful castigation of subaltern modes of touching animals (Ritvo 1987) in which the cruel treatment of animals was taken as evidence of personal and wider cultural moral failing. In response Baker (1854) sought to reassure the domestic audience of his popular writings by differentiating his hunting from 'barbaric' local practices and emphasising the dignity of his elephants' deaths.

These historic tensions between modalities for touching and being touched by elephants constitute but one (albeit spectacular) chapter in a long spatial history that haunts the performance and affections of contemporary Western visitors to Sri Lanka. For example, in a different incarnation Western volunteers travel to

Sri Lanka as 'scientific ecotourists' to observe and research elephants in national parks (Lorimer 2009). In encounters framed by the assemblages and territories of colonial natural history they are encouraged by their natural scientist guides to emulate Tennent; to quiet their passions to bear impartial witness to elephant presence, behaviour and ecology. Through the disciplinary force of visual technologies like binoculars, GPS and photography they are to patiently identify and aggregate individual animals to produce a species population. However, many of these tourists reported that they were travelling for experience and touching encounters; they expected affective extremes – risk, adventure and the unusual. Waiting, watching, counting and entering data was unsatisfactory; good science requires too many absences and zeros. In response, and often against their better judgement, pressurised scientists sought to engage with the animals themselves. They hunt elephants on foot or in jeeps, shooting them at close range with cameras and triggering angry combative responses from territorial males.

This displacement of vision and the desire for corporeal reciprocal encounters with wild animals is more clearly expressed in the relationships sought out by Amy and Sally above and by the thousands of others who pay and travel every year to touch captive animals. This is a new phenomenon worthy of some reflection. One way to make sense of this shift might be to link it with work that identifies the wider and growing prominence that is afforded touch in the sensuous economies of late capitalism (Howes 2005). Here it is argued that the visual logic of modern consumer capitalism has been developed and expanded into multisensory opportunities for embodied consumption. In connected writings on late modern cultures of nature, Thrift (2007) identifies a diverse array of haptic practices within an emerging 'experience economy' (Pine and Gilmore 1999) that aim to allow consumers to 'feel life' – putting them (back) in touch with an assumed authentic natural body. Various modes of nature-based tourism have emerged as key sites for such experiences (Bulbeck 2005, Crouch and Desforges 2003, Curtin 2005). These include adventure tourism – like elephant safaris – which accentuates visceral experience, risk and awe (Cloke and Perkins 2002). While the therapeutic encounters between individual elephants and female volunteers at the elephant sanctuary express a different affective logic, redolent of the growing popularity of various modalities of health tourism and recreation (Lea 2008, McCormack 2003, Paterson 2005).

Amy and Sally's enthusiasms for organised touching encounters with elephants are symptomatic of a growing belief – amongst both popular and scientific audiences – in the therapeutic and recuperative potentials of certain forms of human-animal encounter. For example, in work documenting the enormous growth of opportunities to touch and swim with dolphins, Bryld and Lykke (2000) argue that for postmodern enthusiasts dolphins figure as spiritual – even telepathic – guides to 'a simple, true and sacred life in harmony with nature' (quoted in Bulbeck 2005, 131). They suggest that consumers of dolphin encounters seek to escape from the alienation of modern life to re-enchant a naturalised bond. The sense of touch is primary here as an authentic mode of interspecies communication (Curtin and Wilkes 2007). Dolphins – alongside a range of more mundane companion animals

– have been used to treat children and adults with depression and neurological illnesses (Franklin et al 2007, Marino and Lilienfeld 2007). This animal assisted therapy is the subject of a burgeoning (though) contested scientific literature (Fine 2006). Elephants are less amenable to such encounters than dolphins, dogs and cats, but it is clear from Sally and Amy's explanations that they are valued for their recuperative potential.

It might also be possible to link the rise of touching environmentalisms with both the progressive feminisation and the gender politics of Western environmentalism. In Sri Lanka, for example, female visitors and volunteers outnumber men at touching elephant attractions and this pattern has been identified at captive animal attractions elsewhere (Bulbeck 2005). While Western research on elephant ethology and welfare has been led by a small number of female scientists – including Daphne Sheldrick, Joyce Poole (1996) and Cynthia Moss (2000). Following in the footsteps of female primatologists (Haraway 1989), these women with compassion have challenged the power and philosophies of the men of vision who dominate wildlife conservation (Thompson 2002). Their work identifies elephants as emotional individuals, living in complex family units and susceptible to great personal and social trauma (Poole and Moss 2008). One should be wary of gendering empathy here – Tennant (1867) showed great affection towards captive elephants, even raising an orphaned baby elephant in his house, hand feeding it at the dinner table – but it would be interesting in future work to interpret the rise of touching environmentalisms through recent interventions by poststructuralist feminists that place touch and the body at the centre of critical analysis (Diprose 2002, Grosz 1994, Haraway 2008). This would require data that was not collected in this study.

The material traces of very different spatial histories are animated within the other bodies and relations that comprise our interspecies triangle. The history of Sri Lankan mahoutship with its array of haptic skills and modes of interspecies communication is not well recorded and would offer a fascinating arena for future research. It is known that elephants have been captured and trained on the island for more the 4000 years and that Sri Lanka was once a key node in the international networks for elephant trade and training (Jayewardene 1994b). Parallel work by anthropologists on cultures of horse companionship has identified significant historical and geographical variations in modes of touch (Birke and Brandt 2009) and the same may well be true for elephants. The swift contemporary decline in global captive elephant populations and associated opportunities for traditional mahouts makes this research especially important as there is a real risk that the unrecorded, embodied knowledges of generations of mahouts will die out (Lair 1997).

In spite of the popularity of elephants as academic research subjects, there is also little work that traces the historical geographies of touch and companionship from a pachyderm perspective. As I have explored elsewhere (Lorimer 2010b), the genetics, bodies, ecologies and ethologies of Sri Lankan elephants are haunted by spatial histories as complex and disparate as those of the Western volunteers who travel to touch them. For example, Somalia may well be descended from elephants originating in what is now Burma that were imported to the island as part of a long history of trade and subsequently went feral (Vidya et al 2009). Similarly, at a microbial level Somalia may well harbour tuberculosis, a zoonotic virus that has evolved in the spaces between human and animal bodies and which poses risks to wild and captive elephant populations as anti-biotic resistant varieties proliferate along the increasingly global networks of ecotourism (Alexander et al 2002). Here human and animal bodies touch, intermingle and are disorganised at prepersonal scales with important analytic and policy implications (see Lorimer 2010b).

Touching Conflicts

These spatial histories frame ongoing and often vitriolic debates about the best way of touching and being touched by Asian elephants. These stir up postcolonial politics and contested animal ethics and relate to the nature and legitimacy of elephant captivity in the emerging affective economies of contemporary ecotourism. The captivity paradox – that captivating encounters with wild animals like elephants, dolphins and lions frequently require that these animals are kept in captivity – pits an assertive and well-funded animal welfare lobby against captive elephant owners and managers in both the developed and the developing world. Much of this debate centres on the ethics of the 'free contact' method for elephant management associated with traditional practices of mahoutship. This issue provides an important arena for exploring the politics of touch that is associated with the rise of touching environmentalisms. It also provides a compelling illustration of both the possibilities and the challenges of critically intervening to ensure human-nonhuman conviviality in touching encounters.

In 2002 the RSPCA funded extensive research into the welfare of captive elephants (Clubb and Mason 2003). Drawing on the published literature on elephant behaviour the authors identify respect, rather than fear as the most important currency in the functioning of elephant social groups. They contest claims that physical dominance and corporeal chastisement are normal components of the social order of free-ranging elephants. This intervention challenges the naturalised use of pain for breaking and conditioning elephants and draws attention to the debilitating effects of physical punishment on the animals' mental health. In contrast to the free contact model they review – and advocate on behalf of – forms of 'protected contact' and 'hands-off' elephant management that have been developed and refined in Western zoos (Desmond and Laule 1991). In this regime the elephant handler is physically protected from the animal by bars and cages. Elephants are trained through 'positive reinforcement' or 'operant conditioning' techniques – first pioneered with horses and dogs – which reward good behaviour and do not use physical punishment. In theory, young elephants do not need to be broken and are allowed to act voluntarily in response to oral commands; total obedience is not necessary.

When challenged Sri Lankan defenders of the free contact model argued that separating handlers from their elephants may have negative consequences for the health of the animals. They suggested that trained elephants are happy doing varied physical work when the requirements are clear and that elephants managed under free contact are able to range more widely within captive settings and thus get more exercise and stimulation. They claimed that elephants come to view their handlers as members of their social group and can enjoy interactions with people. Companion handlers increase the elephants' perceived herd size and can offset some of the social and behavioural problems associated with small groups in captivity. These interspecies contacts and connections are severed in a protected model where the mahout can not touch and interact with his elephant (see Clubb and Mason 2003 for a reasonably even-handed review of these arguments). Meanwhile local conservationists and politicians defend the cultural importance of captive elephants – as catalysts for wild elephant conservation and for the vital role they play in religious ceremonies (Fernando 2000).

These debates over modes of touching elephants illustrate frictions (Tsing 2005) between the contrasting modes of 'professional vision' (Goodwin 1994) – or more accurately expert embodied knowledge (Lorimer 2008) – of both traditional elephant handlers and animal welfare scientists and campaigners. In very different ways both groups have developed embodied methodologies for learning to be affected (Despret 2004) – or getting in touch – with elephant moods and behaviours. Both employ modes of critical anthropomorphism in which they filter their knowledge of the animal through specific ethical frameworks to decide how elephants should best be treated. I am not in a position to arbitrate between these approaches but it is clear that the articulated contrasts invoke discourses that have long and undistinguished spatial histories. For example, captive elephant owners and managers in Sri Lanka and Thailand have been targeted in campaigns by Western animal welfare organisations and their local partners (Born Free 2006, PETA 2010). At the shrill end these highlight the 'savage' use of the ankus and make tacit accusations of 'barbarity', arguing that such practices are out of step with the compassionate, civilised and global(ising) norms of Western animal welfare. These arguments sound postcolonial echoes of the historic discourses mobilised by the RSCPA against subaltern modalities of touching animals and undermine efforts to develop modes of cosmopolitan environmentalism.

Meanwhile a long-standing reformist movement has concentrated on developing and implementing technologies for enriching the environments of captive elephants to reduce physical and mental health problems. On a basic level these aim to increase the size of both the elephants' enclosures and of the captive herd (Rees 2009). To design places that reduce elephant boredom these interventions also attune to the distinctive qualities of the pachyderm sensorium. For example, haptic 'feeder innovations' require elephants to work for their food over relatively long time periods (Stoinski et al 2000), while experiments with objects, smells and sounds give curious animals diverting stimuli for trunk and mouth based explorations (Wells 2009, Wells and Irwin 2008, Wiedenmayer

1998). Comfortable substrates have been designed on which captive elephants can scratch, bath and wallow to maintain the health of their feet and skin (Meller et al 2007). Pioneering work with orphaned African (Bradshaw 2009) and Asian elephants (Jayawardena et al 2002) has demonstrated the therapeutic potential of human touch in rehabilitating and re-socialising animals suffering from forms of post-traumatic shock resulting from the death of family members. Here experienced handlers communicate with young elephants through voice and touch in an attempt to re-socialise them for a return to the wild.

If we exclude the romantic and the anthropocentric identities offered by the extremes in these debates – which cast the elephant as either a tool or as an wild beast – then we can identify shared concerns for how best humans and elephant might live together in modes of touching environmentalism that entail convivial forms of human-animal companionship. Here elephants are understood as neither truly wild nor subservient domesticates but instead as companion species (Hutchins 2006). The subsequent debate concerns the nature of the relations that constitute this companionship and the economies they support. As I have argued elsewhere (Lorimer 2010b), reconciling the claims of different human and elephant protagonists will require forms of pluralism (Connolly 2005) or 'cosmopolitics' (Hinchliffe et al 2005) that take seriously more-than-human bodies, senses, knowledges and livelihoods. We can identify some of the challenges for developing these approaches in the tone and the content of the conflicts summarised above.

Conclusions

In this chapter I have identified and illustrated the importance of touch in human-animal interactions. This narrative features diverse modalities, including the therapeutic touch of Sally, the lethal sporting touch of Samuel Baker, the disciplinary touch of elephant breakers, the empathetic touch of Thilak and the curious touch of Somalia. While touch has always been central to the production of natural knowledge and in catalysing concerns for nonhuman others, here I have argued – along with others – that there is a growing willingness amongst Western environmentalists and ecotourists to articulate and undertake touching encounters. This chapter has focused on one compelling example of these touching environmentalisms to trace the different modalities of touch performed in encounters between Western visitors, mahouts and elephants in Sri Lanka. It has argued that we need to situate these modalities in their complex and diverse spatial histories to understand the character, causes and consequences of the emergence of touching environmentalisms. Here I link this rise to the postmodern experience economy, the popularity of animal therapy and the feminisation of environmentalism. This is a speculative analysis in need of future work. I have argued that to understand the politics of touch that characterises contemporary debates relating to the captivity paradox we need to be aware of the colonial histories by which it is haunted and the convivial imperatives it broaches. In conclusion I want to briefly draw out some general observations of what environmentalists might learn from attending to touch and to identity some avenues for future research.

First, tuning in to touch helps us to anticipate some of the likely consequences of the changing political economies of contemporary environmentalism that were identified in the introduction. The shift in the source of conservation funding away the state towards non-governmental organisations and private companies has increased the influence of individual citizen-consumers on the scope and operations of conservation. With the rise of the experience economy and its mediated landscapes of desire and concern there is a growing need for conservationists to target touching charismatic organisms and to provide affectively-charged experiences to draw in resources. These emerging economies will have significant consequences for the affective logic and scope of environmental concern and thus the types of organism and landscape that they produce (Lorimer 2009). Some places and animals are just too difficult or undesirable to touch and may be ignored. Meanwhile the demand for encounters with elephants, dolphins and other captivating animals has produced a lucrative market for those supplying captive animals and offering touching encounters. This demand may have significant consequences for wild animal populations and the experience of captive animal stock. In spite of a wealth of case studies (e.g., Bulbeck 2005, Davis 1997), little is known about the scale, operations and economics of this industry and it is in pressing need of further research.

This chapter has sought to critically examine the modalities of touch that characterise captive animal encounters. It is clear that captive elephants are charismatic companion animals, whose peculiar anatomy and sagacious ethology pose difficult but compelling responsibilities on those charged with their care. Elephants are poorly served by ethical frameworks that encourage instrumental dominance, the smothering claustrophobia of empathy or the misanthropic wilderness of the sublime. Convivial and cosmopolitan approaches for human-elephant flourishing must recognise how elephants are touched by humans and how, in turn, elephants touch us. Touching environmentalisms – that take seriously more-than-human forms and knowledges – provide some tentative steps towards respecting these imperatives. The wider challenge is to respect the ethical, political and territorial logics of this model of companionship in the densely populated and contested landscapes that characterise the range of animals like the Asian elephant. It is difficult to imagine how, when it comes to elephants, some form of constraint or captivity can be avoided – even within extensive national parks. The challenge is to develop techniques and territories for the mutual benefit of subaltern citizens and animals where the dominant modalities of touch are agonistic rather than antagonistic.

Finally, this analysis of the role of touch in debates over elephant management foregrounds the vital role of the embodiment in the derivation and contestation of natural knowledge. Taking the mind out of its modern vat, to use Latour's (2004a) phrasing, and exploring the processes through which people and animals learn to be affected by each other and their wider environments raises important epistemological questions about the grounds for knowledge. Diverse and contested forms of knowledge circulate in this story that supplement and exceed the ocularcentrism of powerful models of environmental science. This account has given a brief feel for these embodied ways of knowing – including the immersive and therapeutic affections of conservation volunteers, the practices of mahoutship and the painful conditioning of elephants. There is much more work to do to develop interdisciplinary methodologies for witnessing and evoking the diverse human and nonhuman affects and knowledges that circulate in these encounters. Although these knowledges matter, they currently escape the narrow confines of the formal mechanisms through which the world is made present by powerful environmentalisms (Latour 2004b). There is a great scope for innovative forms of scholarly praxis.

References

Adams, W.M. and Mulligan, M. 2003. *Decolonizing Nature: Strategies for Conservation in a Post-colonial Era*. London: Earthscan.

Alexander, K.A., Pleydell, E., Williams, M.C., Lane, E.P., Nyange, J.F.C. and Michel, A.L. 2002. Mycobacterium tuberculosis: An emerging disease of free-ranging wildlife. *Emerging Infectious Diseases*, 8, 598–601.

Animal Studies Group. 2006. *Killing Animals*. Urbana, IL: University of Illinois Press.

Ansell-Pearson, K. 1999. *Germinal Life: The Difference and Repetition of Deleuze*. New York: Routledge.

Baker, S. 2003. *Sloughing the human, in Zootologies: The Question of the Animal*, edited by C. Woolf. Minneapolis, MN: University of Minnesota Press, 147–64.

Baker, S.W. 1854. *The Rifle and the Hound in Ceylon*. London: Longman.

Barker, J.M. 2009. *The Tactile Eye: Touch and the Cinematic Experience*. Berkeley, CA: University of California Press.

Bekoff, M. 2002. *Minding Animals: Awareness, Emotions, and Heart*. New York: Oxford University Press.

Bennett, J. 2001. *The Enchantment of Modern Life: Attachments, Crossings, and Ethics*. Princeton, NJ: Princeton University Press.

Birke, L. and Brandt, K. 2009. Mutual corporeality: Gender and human/horse relationships. *Women's Studies International Forum*, 32, 189–97.

Born Free. 2006. Pinnewala Elephant 'Orphanage' Born Free Position and Recommendations, <www.bornfree.org.uk>.

Bradshaw, G.A. 2009. *Elephants on the Edge: What Animals Teach us About Humanity*. New Haven, CT: Yale University Press.

Brockington, D. 2009. *Celebrity and the Environment: Fame, Wealth and Power in Conservation*. London: Zed Books.

Brockington, D., Duffy, R. and Igoe, J. 2008. *Nature Unbound: Conservation, Capitalism and the Future of Protected Areas*. London: Earthscan.

Bryld, M. and Lykke, N. 2000. *Cosmodolphins: Feminist Cultural Studies of Technology, Animals, and the Sacred*. New York: Zed Books.

Bulbeck, C. 2005. *Facing the Wild: Ecotourism, Conservation, and Animal Encounters*. London: Earthscan.

Carter, S. and McCormack, D.P. 2006. *Film, geopolitics and the affective logics of intervention*. Political Geography, 25, 228–45.

Cloke, P. and Perkins, H.C. 2002. Commodification and adventure in New Zealand tourism. *Current Issues in Tourism*, 5, 521–49.

Cloke, P. and Perkins, H.C. 2005. Cetacean performance and tourism in Kaikoura, New Zealand. *Environment and Planning D: Society and Space*, 23, 903–24.

Clubb, R. and Mason, G. 2003. *A Review of the Welfare of Elephants in European Zoos*. Horsham, England: RSPCA.

Clutton-Brock, J. 1999. *A Natural History of Domesticated Mammals*. Cambridge: Cambridge University Press.

Colebrook, C. 2005. Derrida, Deleuze and haptic aesthetics. *Derrida Today*, 2, 22–43.

Collingham, E.M. 2001. *Imperial Bodies: The Physical Experience of the Raj, c. 1800–1947*. Cambridge: Polity Press.

Connolly, W.E. 2005. *Pluralism*. Durham, NC: Duke University Press.

Cousins, J.A., Evans, J. and Sadler, J.P. 2009. 'I've paid to observe lions, not map roads!' – An emotional journey with conservation volunteers in South Africa. *Geoforum*, 40, 1069–80.

Crouch, D. and Desforges, L. 2003. The sensuous in the tourist encounter. Introduction: The power of the body in tourist studies. *Tourist Studies*, 3, 5–22.

Curtin, S. 2005. Nature, wild animals and tourism: An experiential view. *Journal of Ecotourism*, 4 1–15.

Curtin, S. and Wilkes, K. 2007. Swimming with captive dolphins: current debates and post-experience dissonance. *International Journal of Tourism Research*, 9, 131–46.

Davis, S.G. 1997. *Spectacular Nature: Corporate Culture and the Sea World Experience*. Berkeley, CA: University of California Press.

Derrida, J. 2005. *On Touching, Jean-Luc Nancy*. Stanford, CA: Stanford University Press.

DeSilvey, C. 2007. Salvage memory: constellating material histories on a hardscrabble homestead. *Cultural Geographies*, 14, 401–24.

Desmond, T. and Laule G. 1991. *Protected Contact Elephant Training Active Environments*, <http://activeenvironments.org>.

Despret, V. 2004. The body we care for: figures of anthropo-zoo-genesis. *Body & Society*, 10, 111–34.

Diprose, R. 2002. *Corporeal Generosity: On Giving with Nietzsche*, Merleau-Ponty, and Levinas. Albany, NY: State University of New York Press.

Doel, M.A. and Clarke D.B. 2007. Afterimages. *Environment and Planning D: Society and Space*, 25, 890–910.

Duffy, R. 2002. *A Trip Too Far: Ecotourism, Politics, and Exploitation*. London: Earthscan.

Elzanowski, A. and Sergiel A. 2006. Stereotypic behavior of a female Asiatic elephant (Elephas maximus) in a zoo. *Journal of Applied Animal Welfare Science*, 9 223–32.

Fernando, P. 2000. Elephants in Sri Lanka: past, present and future. *Loris*, 22, 38–44.

Fine, A.H. 2006. *Handbook on Animal-Assisted Therapy: Theoretical Foundations and Guidelines for Practice*. Amsterdam: Elsevier.

Franklin, A., Emmision, M., Haraway, D. and Travers, M. 2007. Investigating the therapeutic benefits of companion animals: problems and challenges. *Qualitative Sociology Review*, III, 42–58.

Fullagar, S. 2000. Desiring nature: identity and becoming in narratives of travel. *Cultural Values*, 4, 58–76.

Game, A. 2001. Riding: Embodying the Centaur. *Body & Society*, 7, 1–12.

Goode, D. 2007. *Playing With My Dog Katie: An Ethnomethodological Study of Dog-Human Interaction*. West Lafayette, IN: Purdue University Press.

Goodwin, C. 1994. Professional vision. *American Anthropologist*, 96, 606–33.

Gray, N.J. and Campbell, L.M. 2007. A decommodified experience? Exploring aesthetic, economic and ethical values for volunteer ecotourism in Costa Rica. *Journal of Sustainable Tourism*, 15, 463–82.

Grosz, E.A. 1994. *Volatile Bodies: Toward a Corporeal Feminism*. Bloomington, IN: Indiana University Press.

Haraway, D.J. 1989. *Primate Visions: Gender, Race, and Nature in the World of Modern Science*. New York: Routledge.

Haraway, D.J. 2008. *When Species Meet*. Minneapolis, MI: University of Minnesota Press.

Hart, L. and Locke P. 2007. Nepali and Indian mahouts and their unique relationships with elephants, in *Encyclopedia of Human-Animal Relations*, edited by M. Bekoff. London: Greenwood Publishing, 510–15.

Hayward, E. 2005. Enfolded vision: Refracting the love life of the octopus. *Octopus: A Visual Studies Journal*, 1, 29–44.

Hinchliffe, S., Kearnes M. B., Degen M. and Whatmore S. 2005. Urban wild things: a cosmopolitical experiment. *Environment and Planning D: Society and Space*, 23, 643–58.

Honey, M. 2008. *Ecotourism and Sustainable Development: Who Owns Paradise?* Washington, DC: Island Press.

Howes, D. 2005. *Hyperaesthesia, or the sensual logic of late capitalism, in Empire of the Senses*, edited by D. Howes. Oxford: Berg, 281–304.

Hutchins, M. 2006. Variation in nature: Its implications for zoo elephant management. *Zoo Biology*, 25, 161–71.

Ingold, T. 2000. *The Perception of the Environment: Essays on Livelihood, Dwelling and Skill*. London: Routledge.

Jayawardena, B., Perera, B. and Prasad, G. 2002. Rehabilitation and release of orphaned elephants back into the wild in Sri Lanka. *Gajah*, 21, 87–88.

Jayewardene, J. 1994a. Elephant drives in Sri Lanka. *Gajah*, 13

Jayewardene, J. 1994b. *The Elephant in Sri Lanka*. Colombo, Sri Lanka: Wildlife Heritage Trust of Sri Lanka.

Jones, O. 2000. (Un)ethical geographies of human-non-human relations: encounters, collectives and spaces, in *Animal Spaces, Beastly Places*, edited by C. Philo and C. Wilbert. London: Routledge, 268–91.

Katugaha, H. 2008. The last kraal in Sri Lanka. *Gajah*, 29, 5–10.

Kontogeorgopoulos, N. 2009. Wildlife tourism in semi-captive settings: A case study of elephant camps in northern Thailand. *Current Issues in Tourism*, 12, 429–49.

Lahiri-Choudhury, D. 1999. *The Great Indian Elephant Book*. New Delhi: Oxford University Press.

Lair, R. 1997. Gone Astray: The Care and Management of the Asian Elephant in Domesticity. FAO Bangkok, <www.fao.org/docrep/005/ac774e/ac774e00.htm>.

Latour, B. 2004a. How to talk about the body? The normative dimension of science studies. *Body & Society*, 10, 205–29.

Latour, B. 2004b. *Politics of Nature: How to Bring the Sciences into Democracy*. Cambridge, MA: Harvard University Press.

Laurier, E., Maze, R. and Lundin, J. 2006. Putting the dog back in the park: Animal and human mind-in-action. *Mind, Culture and Activity*, 13, 2–24.

Lea, J. 2008. Retreating to nature: Rethinking 'therapeutic landscapes'. *Area*, 40, 90–98.

Legg, S. 2009. Of scales, networks and assemblages: The League of Nations apparatus and the scalar sovereignty of the Government of India. *Transactions of the Institute of British Geographers*, 34, 234–53.

Lestel, D., Brunois, F. and Gaunet, F. 2006. Etho-ethnology and ethno-ethology. *Social Science Information*, 45, 155–77.

Lorimer, H. 2006. Herding memories of humans and animals. *Environment and Planning D: Society and Space*, 24, 497–518.

Lorimer, J. 2007. Nonhuman charisma. *Environment and Planning D: Society and Space*, 25, 911–32.

Lorimer, J. 2008. Counting corncrakes: The affective science of the UK corncrake census. *Social Studies of Science*, 38, 377–405.

Lorimer, J. 2009. International volunteering from the UK: What does it contribute? *Oryx*, 43, 1–9.

Lorimer, J. 2010a. Moving image methodologies for more-than-human geographies. *Cultural Geographies*, 17, 237–58.

Lorimer, J. 2010b. Elephants as companion species: the lively biogeographies of Asian elephant conservation in Sri Lanka. *Transactions of the Institute of British Geographers*, 35, 491–506.

Lorimer, J. and Whatmore, S. 2009. After 'the king of beasts': Samuel Baker and the embodied historical geographies of his elephant hunting in mid-19th century Ceylon. *Journal of Historical Geography*, 35, 668–89.

Mackenzie, J. 1997. *The Empire of Nature: Hunting, Conservation, and British Imperialism*. Manchester: Manchester University Press.

Marino, L. and Lilienfeld, S.O. 2007. Dolphin-assisted therapy: More flawed data and more flawed conclusions. *Anthrozoos*, 20, 239–49.

Marks, L.U. 2002. *Touch: Sensuous Theory and Multisensory Media*. Minneapolis, MN: University of Minnesota Press.

May, R.M. 1988. How many species are there on Earth? *Science*, 241, 1441–49.

McCormack, D.P. 2003. An event of geographical ethics in spaces of affect. *Transactions of the Institute of British Geographers*, 28, 488–507.

McCormack, D.P. 2008. Engineering affective atmospheres on the moving geographies of the 1897 Andree expedition. *Cultural Geographies*, 15, 413–30.

Meller, C. L., Croney, C.C. and Shepherdson, D. 2007. Effects of rubberized flooring on Asian elephant behavior in captivity. *Zoo Biology*, 26, 51–61.

Moss, C. 2000. *Elephant Memories: Thirteen Years in the Life of an Elephant Family*. Chicago: University of Chicago Press.

Mullan, B. and Marvin, G. 1987. *Zoo Culture*. London: Weidenfeld & Nicolson.

Paterson, M. 2005. Affecting touch: towards a 'felt' phenomenology of therapeutic touch, in *Emotional Geographies,* edited by L. Bondi, J. Davidson and M. Smith. Aldershot, England: Ashgate, 161–76.

Paterson, M. 2007. *The Senses of Touch: Haptics, Affects, and Technologies*. Oxford: Berg.

PETA. 2010. *Help Thai Elephants*, <www.helpthaielephants.com>.

Pine, B.J. and Gilmore, J.H. 1999. *The Experience Economy: Work is Theatre & Every Business a Stage*. Boston: Harvard Business School Press.

Poole, J. 1996. *Coming of Age with Elephants: A Memoir*. New York: Hyperion.

Poole, J. and Moss C. 2008. Elephant sociality and complexity: the scientific evidence, in *Elephants and Ethics: Toward a Morality of Coexistence,* edited by C. Wemmer and C. Christen. New York: Johns Hopkins University Press, 69–98.

Pratt, M.L. 1992. *Imperial Eyes: Travel Writing and Transculturation*. London: Routledge.

Rees, P.A. 2009. The sizes of elephant groups in zoos: Implications for elephant welfare. *Journal of Applied Animal Welfare Science*, 12, 44–60.

Ritvo, H. 1987. *The Animal Estate: The English and Other Creatures in the Victorian Age*. Cambridge, MA: Harvard University Press.

Rodriguez, J.P., Taber, A.B., Daszak, P., Sukumar, R., Valladares-Padua, C., Padua S., Aguirre, L.F., Medellin, R.A., Acosta, M., Aguirre, A.A., Bonacic, C., Bordino, P., Bruschini J., Buchori, D., Gonzalez, S., Mathew, T., Mendez, M., Mugica, L., Pacheco L. F., Dobson, A.P. and Pearl, M. 2007. Environment – Globalization of conservation: a view from the South. *Science*, 317, 755–56.

Rothfels, N. 2008. The eyes of elephants: changing perceptions. *Tidsskrift for Kulturforskning*, 7, 39–50.

Sanderson, G. 1896. *Thirteen Years Amongst the Wild Beasts of India*. London: WH Allen & Co.

Scigliano, E. 2004. *Love, War and Circuses: The Age Old Relationship Between Elephants and Humans*. London: Bloomsbury.

Sobchack, V.C. 2004. *Carnal Thoughts: Embodiment and Moving Image Culture*. Berkeley, CA: University of California Press.

Stoinski, T.S., Daniel, E. and Maple, T.L. 2000. A preliminary study of the behavioral effects of feeding enrichment on African elephants. *Zoo Biology*, 19, 485–93.

Sukumar, R. 2003. *The Living Elephants: Evolutionary Ecology, Behavior, and Conservation*. New York: Oxford University Press.

Tennent, J.E. 1867. *The Wild Elephant and the Method of Capturing and Taming it in Ceylon*. London: Longmans.

Thompson, C. 2002. When elephants stand for competing philosophies of nature: Amboseli National Park in Kenya, in *Complexities: Social Studies of Knowledge Practices,* edited by J. Law and A. Mol. Durham, NC: Duke University Press, 166–90.

Thrift, N. 2007. *Non-representational Theory: Space, Politics, Affect*. London: Routledge.

Tilley, C.Y. 2004. *The Materiality of Stone*. Oxford: Berg.

Tsing, A.L. 2005. *Friction: An Ethnography of Global Connection*. Princeton, NJ: Princeton University Press.

Turner, J. 1980. *Reckoning with the Beast: Animals, Pain, and Humanity in the Victorian Mind*. Baltimore, MD: Johns Hopkins University.

Varner, G. 2008. Personhood, memory and elephant management, in *Elephants and Ethics: Toward a Morality of Coexistence,* edited by C. Wemmer, C. Christen and J. Seidensticker. Baltimore, MD: Johns Hopkins University Press.

Vidya, T.N., Sukumar R. and Melnick D.J. 2009. Range-wide mtDNA phylogeography yields insights into the origins of Asian elephants. *Proceedings of the Royal Society – Biological Sciences*, 276, 893–902.

Wells, D.L. 2009. Sensory stimulation as environmental enrichment for captive animals: A review. *Applied Animal Behaviour Science*, 118, 1–11.

Wells, D.L. and Irwin, R.M. 2008. Auditory stimulation as enrichment for zoo-housed Asian elephants (Elephas maximus). *Animal Welfare*, 17, 335–40.

Wemmer, C.M. and Christen, C.A. 2008. *Elephants and Ethics: Toward a Morality of Coexistence*. Baltimore, MD: Johns Hopkins University Press.

Whatmore, S. 2006. Materialist returns: practising cultural geography in and for a more-than-human world. *Cultural Geographies*, 13, 600–09.

Whatmore, S. and Thorne L. 2000. Elephants on the move: spatial formations of wildlife exchange. *Environment and Planning D: Society and Space*, 18, 185–203.

Wiedenmayer, C. 1998. Food hiding and enrichment in captive Asian elephants. *Applied Animal Behaviour Science*, 56, 77–82.

Williams, J.H. 1950. *Elephant Bill*. Harmondsworth, England: Penguin.

Yusoff, K. 2007. Antarctic exposure: archives of the feeling body. *Cultural Geographies*, 14, 211–33.

Chapter 9

Towards Touch-free Spaces: Sensors, Software and the Automatic Production of Shared Public Toilets

Martin Dodge and Rob Kitchin

Introduction

> The public restroom, so unattended by social scientists, is surely a site of analytic riches. ... tensions form around who we are, what we are to share, and with whom we are to share it. (Molotch 2008, 61)

New software-enabled technologies are changing the social and material production of everyday landscapes, and re-figuring the embodied relationships between people and the environment through touch. The places where people are allowed, obliged and forbidden from touching particular technological objects represent a complex and delicately patterned landscape, but one that is negotiated largely in a habitual, non-conscious fashion. Touching with hands is integral to so much technologic activity and control – the pressing of buttons, pulling of handles, flicking switches, twisting selector dials, and so on. Nearly half the working surface area of the laptop used to compose this chapter is a keyboard and touch-pad ergonomically designed for average human hands to engage with software. And yet touch is an overlooked spatial sense and practice in human geography (although see Hetherington 2003, Paterson 2007, Dixon and Straughan 2010). It is somewhat ironic then that in this chapter we are concerned with the *reverse* situation, as we interrogate the nature of mundane technologies that are designed to work *without* direct human touch.

As such, we consider how tools and appliances are being designed and engineered to interact and respond appropriately to people by remotely sensing the presence of human bodies, and offering modes of control based on proximity rather than actual physical touch (there are other non-tactile approaches to computer control such as sound-activated controls and speech recognition interfaces, but these are beyond the scope of this discussion). We want to focus here on electronic/digital technologies, being applied in everyday contexts, that use sensors and software to automatically produce spaces that can react to people or, at a minimum bodily-shaped objects, in meaningful ways without direct contact. An increasing number of examples are evident in public buildings and office environments, such as software-controlled doors that open automatically when a person approaches, lights and air conditioning that turns itself on when a sensor detects human motion in a room (and which turns itself off again when the space empties), and keyless locks that open with the proximity of contact-less radio frequency identification (RFID) cards. Indeed, digital sensors and decision-making software are all about us, monitoring background infrastructures, supervising utility services, regulating material flows, animating objects and environments, and enrolled in solving the myriad tasks of daily living.

The phenomenal growth and influence of digital technologies on everyday activities is due to the emergent and executable properties of software; how it codifies the world into rules, routines, algorithms, data lists and structured databases, and then executes these to do useful work that changes practices and how spaces come into being (Kitchin and Dodge 2011). While software is not sentient and conscious it can still exhibit some of the characteristics of 'being alive' (Thrift and French 2002, 310). This essence of 'being alive' is significant because it means computer code can make things do work in the world in an autonomous fashion – that is, it can receive inputs from its environment and process this information, make decisions and act on them without human oversight or authorisation. When software executes itself in this automatic way it possesses what Mackenzie (2006) terms 'secondary agency'. However, because software is embedded into familiar objects and enclosed systems in often subtle and opaque ways, its presence and power is little considered and is typically only noticed when it performs incorrectly or fails (cf. Graham 2009).

Recently the role of touch to control software has become much more apparent and, one might argue, more intensively tactile. The conventional keyboard/ mouse input devices are being rapidly supplanted as many of the most desirable and successful handheld consumer technologies, such as mp3 players, satnavs and especially mobile phones, are operated through sophisticated touch-based screen interfaces that are compellingly intimate and intuitive to use. Touch-screen interfaces are now rapidly becoming routine, emplaced within innumerable city and office spaces such as the control panels of photocopiers, vending machines, information kiosks and parking meters. Software is enrolled to bring space into being in particular ways, and increasingly to change where people touch surfaces, how they touch to control things and make objects perform tasks, and conversely how software mitigates the need for touch in certain instances. Yet the effects of software on everyday tactilities has not been documented by social scientists (although see Paterson 2007). Research is therefore needed that can account for the tremendous scale and speed of the growth of code, including within all kinds of mundane service spaces, and to understand the productive capacity that software has to make the world differently in terms of its materiality, economic relations, social processes and everyday practices. This should include those practices most intimately associated with the body, such as toileting.

To begin to explain the nature of this automatic production of touch-free spatiality (after Thrift and French 2002) we concentrate our analysis on shared public toilets, vital but somewhat disregarded spaces of modern life. The focus of the analysis presented here is on 'globalised' Western-style public shared toilets that are the norm in UK and Ireland. We do this while also recognising more globally the wide imbalances of access to any formal toilet facilities, and that lack of basic sanitation remains a major cause of unnecessary deaths, reflecting and reinforcing the uneven geography of development across the world (cf. George 2008, Jewitt 2011).

Bathrooms outside the home are culturally complex spaces, with multiple ambiguous meanings, providing public spaces for very necessary, private activities, but also spaces that are necessarily shared. In using public toilets many people have anxieties around privacy, personal safety and perceived risks of exposure of intimate activities to others and, above all, a sense of vulnerability through enforced sharing of space with strangers (cf. Molotch and Noren 2010). Here we analyse how some toilet spaces are being reshaped, as technologies are applied that seek to render toileting practices into a sequence of touch-free activities, and attempt to diminish direct handling of the materiality of the bathroom surfaces and fixtures. Driven by a range of modernist discourses around hygiene, convenience, and efficiency, it is apparent that many public toilets are now sites of sensors and software deployed to react to humans without direct touching: to flush toilets automatically, to dispense soap and water without touching a lever or turning a tap, and sensing the presence of wet hands waiting for drying. However, the logics of software-enabled automation able to overcome the fear of contamination and subconscious disgust at direct touching of surfaces shared with strange bodies is frequently nullified because the actual deployment of touch-free sensors is typically incomplete and oftentimes haphazard, most evident in the inconsistency and therefore ambiguity involved in walking up to what might or might-not be automatically opening doors. We conclude by considering why the spaces of touch are only ever partially reconfigurable by software technologies, and what this might mean for the automation of other everyday environments and tactile engagements.

Toilet Spaces, Toileting Practices

> People care a great deal how they pee and shit. Their strivings for decency confront the facilities available to them as well as the social strictures and hierarchies that order who goes where. (Molotch 2008, 60)

Daily toileting is an elemental physiological function. It is enveloped in a range of cultural practices and complex social meanings. It is enacted in spaces variously configured to conceal these practices and within architectural forms that reflect and reify these meanings. In Western countries toilets are ubiquitous, found in virtually all dwellings and available to occupants of public buildings in industrialised nations, although their fixtures, materials and layout vary somewhat from place to place (cf. George 2008). For most people in these countries access to specifically designed bathroom spaces, comprising functioning flush water closet (WC) and sink with clean running water, is seen as essential for convenient and comfortable living.

Toilets are at once mundane, but also an essential service space that everyone uses. Despite its ubiquity, toileting in Western cultures is typically constructed as a most private and solitary function, except for young children. Consequently, the toilet is understood as a taboo space because of the 'uncivilised' practices it seeks to conceal from the knowing gaze of others. Understanding the toilet as an ambiguous and taboo space revolves around notions of what is clean and what is dirty. Here, the work of anthropologist Mary Douglas (1966) is useful in explaining that dirty and clean are not innate characteristics, but are culturally constructed categories that arise out of processes of social ordering and the production of normative behaviour. Key to the construction of the category of 'dirty' is that it can be defined as 'matter out of place' ('Shoes are not dirty in themselves, but it is dirty to place them on the dining table', Douglas 1966, 36.) 'Matter out of place' varies with cultural context, but is seen as entirely natural to those living within a given culture. While the symbolic boundaries between categories seem strong, they must be continuously maintained, for example with prohibitions, rules and purity rituals that seek to keep matter in the correct place and to punish those who transgress. The shared public toilet is a troubling space because such boundaries are particularly at risk.

The spatiality of being 'in place/out of place' (Cresswell 1996) can be finely grained, for example in the differentiating boundaries between 'clean' and 'dirty' within a bathroom cubicle or even parts of the WC unit. As Bichard et al. (2008, 81) note: '[t]oileting residue on the toilet seat can be considered dirty as opposed to it being in the toilet bowl; thus a matter of degree can shift our concept of what we consider clean or soiled.' Often matter becomes 'out of place' because of the perceived spatial position of an object relative to 'dirty' activities, and also the physical distance to other surfaces that might be harbouring germs. Something that is initially classified as 'clean' may come too close to (but not actually touch) a 'dirty' object or practice and thus itself become 'dirty'. Maintaining 'matter in place' is not just then the avoidance of direct tactile contact, it is about proximity and notions of acceptable distance. The degree and duration of touch, if it occurs, can also matter. Just a quick touch of a finger tip on a button might be perceived differently from the requirement to give a firm press of a handle with the palm of the hand.

The work of the categorisation of 'dirt' in determining bodily behavioural and social rules rests to a large degree on the notion of disgust. This powerful emotion compels people to avoid the presence and especially direct contact with sites, objects, individuals, activities that are normally classified as 'dirty'. Contact by sight, smell, sound and especially touch with bodily fluids and human wastes, particularly those of strangers, is widely regarded as particularly disgusting (cf. Miller 1998). Excrement, for example, generates an affective response of revulsion and fear. As 'matter out place' it needs to be treated specially – quick disposal that avoids contact with bare hands. Indeed, in a hierarchy of human senses it is touch that can evoke disgust most powerfully because 'matter out of place' might possibly enter the body. As such, touching disgusting things is to be avoided at all costs as it implies possible physical contamination through the skin or by ingestion.

Public toilets are inherently disgusting places because of the unavoidability of physical contact by one's own skin onto surfaces used by others, and consequent fear of contamination from other people's bodily residues (faeces, urine, hair, skin flakes, sweat, saliva/spit, vomit, mucous, blood), both seen and unseen (Greed 2006; Bichard et al. 2008; Molotch and Noren 2010). In shared toilets this can be accompanied by their associated smells, commingling with the background chemical cleaning products, and the sounds of others performing: groans, farts, sputters and plops, and satisfied sighs. One might also on occasion literally feel the presence others: '[w]e all know ... the sensation of a toilet seat still warm from a prior body, the stranger sensed in so disquieting a way' (Molotch 2008, 61). Affective responses to the toilet space are heightened by disturbances to the general sense of orderliness and maintenance which can be invoked by unidentifiable stains on the cubicle walls, grimy looking smears on surfaces, scratches, cracked tiles, vandalism in the form of graffiti, burn marks, and broken fixtures, the presence of litter and loose toilet paper ('matter out of place'). The extent of these signifiers in aggregate can mark a public toilet as uncared for, and thus unclean.

The toilet is then a deeply problematic site, and doubly so when a public facility. It is an arena in which 'matter' from human bodies routinely becomes 'out of place'. Western toilets, with flush WCs, are designed to engender control of such 'matter out of place' as far as possible and to remove it quickly and hygienically. The design and use of technological systems for waste control are also accompanied by particular toilet cleaning regimes to disinfect surfaces, along with the necessity to clear occasional blockages and maintain plumbing in working order. Touch-free technologies, as the latest iteration in bathroom design, resonate with the scalar spatiality of disgust and seek to provide automated mechanisms to maintain bodily distance from potential 'matter out of place'. Although users still might see and smell 'matter out of place', and thus have an awareness of sources of disgust, they are protected against physical contact with it. Touch-free technologies are therefore fundamentally about disgust control, although this is usually dressed up in the more delicate language of hygiene and efficiency (see discussion below).

Toilets Technologies

> [T]he chances of pathogen transmission are very high even in toilets that may appear to look clean, as every door handle (especially the last one out to the street), tap, lever, flush, lock, bar of soap, toilet roll holder, and turnstile, is a potential germ carrier. (Greed 2006, 128)

Even a basic bathroom, in the modern western context, is a highly technological space, reliant on a raft of scientific and engineering developments to make it function as required. Toilets are also tangible contact points between human bodies and the sewer network, a vital but hidden infrastructure to channel, control and remove 'matter out of place'. Toilet technologies need to be efficient in performing hydraulic tasks. While water flows easily with gravity, it is heavy to move and difficult to fully contain, and must be reliably supplied. Many ingenious mechanical solutions have been engineered to safely regulate the supply of water – siphonic cisterns, self activating cut-off valves, overflow outlets – and, in some senses, to automate aspects of toilet space and thereby compensate for human oversight and lassitude. Safety is also a particular issue in terms of heating water and carefully separating water from the electrical equipment. This might partly account for the relative lack of integration of electrical appliances and electronic technologies into bathrooms, particularly in comparison to other domestic and work spaces. In many respects, the technicity of modern plumbing and bathroom fixtures only becomes apparent in failure: a blocked waste pipe reveals just how quickly the convenient sense of a normal flush toilet can unravel (cf. Graham 2009).

A range of plumbing techniques, along with specially designed hygienic materials, are deployed in toilets to increase the psychological detachment from the physiological acts of defecation and thereby to counteract fears of contamination, and they also support ritualistic aspects of cleanliness such as hand washing. Examples include the WC u-bend that holds a reservoir of water to block sewer smells, a powerful flush that whisks away waste, sinks with running water on-demand, the wipe-clean white ceramic tiles that can be easily inspected for (visible) dirt. Technological advances in the name of cleanliness, however, do not necessarily perform unproblematically. As Greed (2006, 129) comments: '[o] stensibly, hygienic equipment, such as electric hand-driers (often imagined to be safer than towels) may blow germs back into the atmosphere.' While surfaces may appear to be clean, there could lurk hidden hygiene problems in toilets, including recent fears of newly resistant 'superbugs', evolved, in part, as a result of anti-bacterial cleaning regimes.

Evolving technological solutions have sought to render shared public toilets ever more automated in recent decades. Automation is presented as advantageous to the users of the toilets and to those who have responsibility for maintaining and managing them. Our primary concern here is with development of digital technologies that are designed to negate the need to touch toilet fixtures. Such automation works, we would argue, because it makes toilet technologies progressively more distanced and opaque in use. For example, operation of the standard flush WC has evolved from the once common pull chord to physically release water from an overhead cistern to a push lever on the side of the WC cistern, and now widespread pressing of duo-flush buttons on top of the cistern offering choice of big and small flows. The latest trend is touch-free flush controlled by waving over a strategically positioned passive infrared (PIR) sensor that activates a control circuit to release a calculated volume of water from a hidden cistern (Figures 9.1 and 9.2), and the next development is no direct human operation at all, where software activates the flush when a sensor detects the user vacating the toilet seat. This automation translates into diminishing kinaesthetic skills needed to operate the WC, and reduces the duration/intensity of hand touch of control surfaces (Table 9.1). It also has fewer external moving parts to be physically

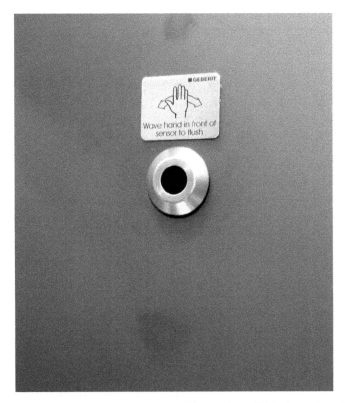

Figure 9.1   A typical 'magic eye' sensor in a WC cubicle in a shared public toilet in the UK. The physical form of the sensor does not follow function hence the presence of the small explanatory sign indicating usage in text and image. The fact that signage is deemed necessary is indicative that these kinds of touch-free sensors are not yet sufficiently common and standardised to be transparent; it is not be necessary to sign the usage of a WC push handle flush

Source: author photograph

manipulated and potentially vandalised. Activities that are harder to automate with touch-free technologies are to do with access in terms of door opening and locking/unlocking, which means the coping practices that Bichard et al. (2008, 80) describe will likely continue:

> …users described how locking the toilet cubicle door could only be done with a handful of toilet paper acting as a barrier between the hand and door lock. This behaviour was considered most beneficial before toileting, to prevent unknown and unseen dirt contaminating the more personal areas of the body.

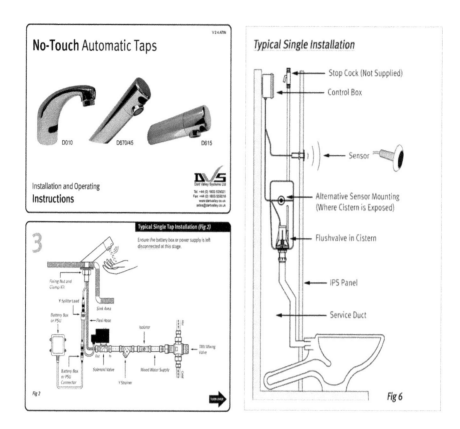

Figure 9.2 **Schematics for typical installation of 'no touch' automatic taps (left hand images) and wave activated WC flush (right hand image)**

Source: Manufacturers pdf brochure, Dart Valley Systems Ltd, <www.dartvalley.co.uk>, 2010

In addition to the WC unit, the most common forms of touch-free bathroom mediation are automatic lighting, taps, hand dryers, urinal flushing, and dispensing of consumables such as toilet paper, soap and towels. Table 9.2 provides a summary of the technologies that are in use in at least some shared public toilets in UK/ Ireland. As discussed below very few, if any, shared public toilets have the full spectrum of automation technology installed.

Crucial to the automation of toileting practices to reduce the sense of disgust are digital sensor technologies. Sensors can operate by detecting changed environmental conditions using different parts of the electromagnetic spectrum including light, sound, heat, as well as the presence of physical material such as smoke, water or human bodies. Such detection has been used routinely in public space, including bathrooms, for many years in alarm systems for fire, flooding and security. Typically they work in a passive way, set up to monitor space and

Table 9.1   The evolving WC technologies in relation to changing levels of direct hand touch of control necessary to complete the task

Flushing a WC toilet	Intensity of tactile contact
Manual sluicing away of waste	Multiple potential hand touches, collecting, aiming and pouring water
Release chain to overhead cistern	Firm grip with whole hand and strong yank
Lever release	Press with fingers or palm of hand
Dual flush button	Light ('fingertip') touch activation
Hand wave PIR sensor	No direct touch, active wave of hand
Occupant / body movement sensor	Passive 'walk away' activation, no conscious interaction to flush or tactile contact

remain inert as long as conditions remain 'normal', only triggering a response if a predetermined threshold level is breached, for example when a high particulate level in the atmosphere sets off the smoke alarm. Having multiple sensors and processing software means location indications can be generated. Sensors are most obvious through separate detector boxes mounted on visible surfaces, but the detector circuits can also be integral to the equipment to monitor its operation (e.g., door opening) and detecting an abnormal operation or failure (e.g., measured water flow indicates the failure of a valve).

Technologies have also offered progressively more control over the toilet space for those responsible for their daily cleaning and general management. For example, hygiene control for urinals, with flushing performed as purely mechanical cycle (cistern fills then flushes, and repeats) systems or via direct activation from the user, have been augmented by electrical controls that offered sequences of flushing and remote activation of 'super flush' for cleaning, for example, and also facilitates removal of direct user activation thereby reducing protruding external fixtures for misuse or vandalism. Updating to electronic systems for urinal flushing meant managers could select different timed flush sequences and also monitor for faults. The addition of sophisticated digital controls with a software interface offers programmable settings and a choice of responses to sensor inputs, as well as logging of performance for later analysis. This is evidence of the shift of local to remote control through sensors and software, and accordingly Braverman (2010, 15) reads this change with Bruno Latour's notion of 'centres of calculation', arguing that:

> [u]nlike the flushometer, which embodies a gaze that is only present in the space of the washroom itself, the central computer manages the washroom from a central location located elsewhere. Hence, the flushing device is not only programmed initially by the manufacturer but through continuous programming and reprogramming.

Table 9.2   Summary of the range of digital technologies available for installation in shared public toilets

Activity	Technology function	Automation / Sensing	Replaces / Augments
User access	Entrance/exit doors	Automatic opening, PIR sensor detects approach of human body	Manual opening with hand; powered-assistance door activated by button press
	WC cubicle door opening/locking	None	Still largely manual opening with hands, mechanical lock
	Access control, fee payment	Electronic opening barriers, digital sensor count people and checks money, software logging of fees and usage statistics	Manual turnstile with mechanical counter
	Lighting	Timed; automated according to daylight; activation in response to human presence via PIR sensor	Always on; electro-mechanical timing; manual activation by light switches
Toileting	Urinal flushing	Programmable settings for variable flush sequences; PIR sensor for flush after use; monitors usage, reports status	Manual activation; electro-mechanical timed flushing
	WC seat cleaning / cover	Activates after flush	Manual cleaning; button push for mechanical dispensing of new cover
	WC paper dispensing	PIR sensor for dispensing of measured amount; potential to monitor usage, reports status	Manual dispensing with hand
	WC flushing	PIR sensor for 'wave' activation and also 'walk away' activation; monitors usage, reports status	Manual activation by hand on lever / button
	Sanitary product bin? [More in female WC?]	None? ??	Manual disposal into sanpro bins
	Accessible WC - Distress alarm / call system [Accessible WC – anything else?]	Digital call circuit routed to control centre; logs usage	Calling for help; electrical alarm trigger and local bell / flashing light to signal attendant
Hand washing	Water dispensing	PIR sensor for touch-free activation; automatic cleaning cycles; monitors use and failsafe cut-off	Manual activation by hand using twist or percussion push taps
	Soap dispensing	PIR sensor for touch-free activation	Soap blocks; manual push button dispensing of liquid soap
	Hand drying – air dryers	PIR sensor for touch-free activation; monitors	Paper towels / roller linen towel; manual

Category	Item		
		usage	activation of dryer by push button
	Hand drying – paper towels	PIR sensor for touch-free dispensing of measured amount; monitors usage	Manual dispensing by hand touch
Environmental and hygiene control	Flushing (complete system for cleaners)	Simultaneous flushing cycle of all units, super flush for deep hygiene clean; monitors use and failsafe cut-off	Manual flushing of units separately
	Heating, AC, ventilation	Programmable and flexible settings. PIR sensors for activation only when space is in use; reports status and logs operation	Manual controls; electronic timings and thermostatic sensors
	Odour control systems (Ozone generator, perfume spray)	Programmable and flexible settings. PIR sensors for activation only when space is in use; reports status	Electromechanical operation, electronic timing
	Air sanitiser ??	??	
	CCTV	Networked, digital system to remote centralised control; logging; potential for algorithmic detection of unusual behaviours	Presence of human attendant onsite; analogue television monitored locally
	Metering of usage	Digital meters, logging status, remote reading, detecting and reporting faults	Mechanical meters, manual reading
	Alarms (smoke, fire, flooding, burglary)	Integrated with BMS, networked for remote monitoring, logging status, reporting failures	Electromechanical alarm linked to bells and lights; electronic alarm operating locally
Miscellaneous	Cleaner time & attendance system	RFID identification, reports failure, logs status	Paper based recording; electronic 'punch cards'
	Vending machines?	Monitors stock level and networked to report status and faults	Periodic restocking

The ultimate degree of automation for management control is in a sense realised by the automated public toilet, typically a free-standing single-user WC unit in the street that requires payment to use. Usage is time limited and they are fully cleaned automatically after each cycle (cf. Braverman 2010).

Promotional Discourses for Automated Toilet Technologies

An examination of the marketing literature of UK toilet technology manufacturers reveals that a wide range of narratives are used to promote touch-free bathrooms that encompass and extend beyond ideas of disgust and 'matter out of place'. For many manufacturers the addition of sensors and software is a significant means of 'adding value' to existing product ranges, to facilitate further sales and/or more profitable pricing structures. Six discourses predominate:

- perceived hygiene and potentially real health benefits
- additional convenience and comfort
- being 'modern'
- easy installation and greater reliability of operation
- enhanced control and configurability
- promise of saving and efficiencies

The operationalisation of these discourses is well illustrated by the promotional brochure for typical automatic taps (Figure 9.3). This brochure encapsulates several of the master narratives around such toilet technologies when it states: "DVS No-Touch products allow you to control your water efficiently, conserve energy and cut down on your costs without sacrificing performance and reliability". Here is the classic 'win-win' technology sales pitch: to be more efficient, but still provide the same service. The stress is also on the control afforded, along with claims of reliability. The key visual element in the advertisement is the automatic taps in operation washing (already clean) hands, accompanied by the claim "Save Water – Improve Hygiene", linking two distinct discourses underlying toilet automation to mutually reinforce each other.

The appeal to saving resources through efficiency is key, with claims that automation offered by sensors and software can deliver significant reductions in water usage: "Up to 65% savings on water costs" (Figure 9.3). Automated taps programmed to supply an 'optimal' burst of water only when hands are directly under the faucet use less water for each cleaning cycle than twist or push taps (Figure 9.4). In a domestic context in UK/Ireland water has typically been supplied unmetered (flat rate annual charging), so there has been little concern with the efficiency of home toilet facilities, but clearly for large institutions with multiple bathrooms in intensive use the charges for water usage are a variable cost that needs to be controlled and ideally reduced. This is doubly so for the costly provision of heated water for hand washing.

Water
Management
Systems

No-Touch
Automatic Taps

No-Touch automatic taps and accessories from Dart Valley Systems
incorporate state-of-the-art design and technology and also offer
hygienic water dispensing solutions to all market sectors - from
hotels to supermarkets, laboratories to hospitals, schools to
universities.

Our extensive product range is suitable for use by the elderly or
disabled and the superior heavy duty construction offers
resistance to vandalism and misuse.

Easy to install, with options for Mains or battery operated, DVS
No-Touch products allow you to control your water efficiently,
conserve energy and cut down on your costs without sacrificing
performance and reliability.

Features

- Up to 65% Savings on Water Costs
- No-Touch Operation
- Hygienic - Helps Avoid Cross Infection
- Easy to Use - Ideal for Disabled or Elderly
- High Performance & Reliability
- Battery or Mains Powered
- Easy to Install & Maintain
- Additional Control Systems Allow Custom Run-Times
  (this option is not available to all models)

D010
Classic Tap

D670/45
Aquarius
DM-A45 Tap

Classic Tap in Situation

Save Water
Improve Hygiene

WRAS

Figure 9.3 **A sample page of a sales brochure promoting the virtues of automatic taps for shared public toilets. The layout, typography and ordering of items in the bullet-point list is revealing of the prioritisation of discourses**

Source: Dart Valley Systems Ltd, <www.dartvalley.co.uk>, 2010

Estimated Water Consumption

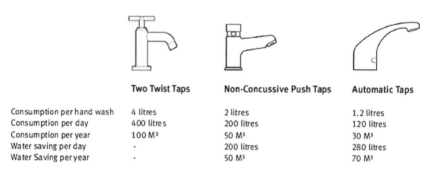

	Two Twist Taps	Non-Concussive Push Taps	Automatic Taps
Consumption per hand wash	4 litres	2 litres	1.2 litres
Consumption per day	400 litres	200 litres	120 litres
Consumption per year	100 M³	50 M³	30 M³
Water saving per day	.	200 litres	280 litres
Water Saving per year	.	50 M³	70 M³

Figure 9.4 Part of the marketing literature for automatic taps is a comparative chart for potential water savings from updating to no-touch taps over conventional faucets

Source: Dart Valley Systems Ltd, <www.dartvalley.co.uk>, 2010

A contemporary subset of the efficiency discourse in promoting technologies is the appeal to sustainability of operations in addition to cost savings: "saving water is good for the environment" (Figure 9.3). Being seen to be 'sustainable' has become a key benchmark for many institutions and corporations, speaking to notions of morality and care for the community. Saving water is one of leading mantras in sustainability, given its iconic status as an essential element for living and its material scarcity in many parts of the world. The automation of toilets can therefore be justified as a sustainable 'solution', especially when it is supported by economic rationality.

For building owners and those responsible for managing shared public toilets the appeal to reliability is another powerful discourse. For any technology subject to intensive usage, it must work as intended day in, day out, with minimal care and maintenance. Shared public toilets have long been notorious as sites for malicious damage and bathroom fixtures must be designed in consequence, with marketing claims such as "superior heavy duty construction offers resistance to vandalism and misuse" (Figure 9.3). Here, the benefits notionally flowing out of new toilet technologies are not around touch-free automation per se but, according to British Toilet Association's 'best practice guide' (BTA 2010, 30): 'A non-touch system with a concealed cistern provides less opportunity to vandalise the unit and is more hygienic.' In a larger sense, reliability is also bound up with issues of installation and maintenance that are stressed as being 'easy' and 'problem-free' (Figure 9.3). Such a prosaic appeal should not be dismissed. Given that some touch-free technologies are still relatively new, the stress is on how manufacturers can offer 'complete solutions' and ones that can be straightforwardly retrofitted into existing toilet spaces.

Another discourse used to promote toilet technologies is control over the space and new means of knowing for building services managers tied to issues of enhancing safety/security, which has become a fundamental promotional discourse in a risk-conscious world. Control is coupled with a configurability that promises greater flexibility for cleaning operations. The programmability through software means it is possible to change parameters to suit local contexts rather than rely on factory defaults often locked into an electronic system. For example, in Figure 9.3 the advertisement lists the feature of "Additional control systems allow custom run-times", indicating that manufacturers believe some customers will pay more for perceived greater degree of control. Managers can also be offered options to override and lock-out water supply to forestall abuse and better cope with vandalism.

Other promotional narratives for 'touch-free' technologies, while aimed at facilities managers, also stress advantages to patrons, detailing how new toilet fixtures work better than existing ones. Discourses around new technologies often claim enhanced convenience in tackling existing tasks or wholly new kinds of tasks, elemental to claims of being modern. Such promises of convenience are central to consumer-oriented societies, with each new round of technology assertively claiming to be easier to use than the preceding ones, reducing the time burden to complete mundane tasks and the cognitive effort involved in sustaining everyday living. Convenience is often stressed for target groups of people who might have suffered from the poor design or operation of existing technologies. As Figure 9.3 notes: "Easy to use – ideal for disabled and elderly". Other manufacturers stress the compliance with disability equality legislation for their automatic toilet products. This kind of claim emphasising the positive attributions of being 'touch-free' however presumes that 'elderly' or 'disabled' are meaningful categories of users, all sharing the same bodily (in)capacities. Research has disputed this, showing how some new automation technologies can make toileting harder in some contexts for some users (cf. Bichard et al. 2006, 2008).

In many respects these discourses represent a continuation of an established but questionable progressive-modernist narrative that technologies can make life better, updated in contemporary contexts in terms of 'digital dreams' and the bold claims for so-called 'smart systems'. Bathrooms, with their specialised equipment and fittings, have long been sold as sites of modernity and a place for displaying one's tastes and distinctions in terms of consumption. Modern technologies are promoted through their capacities to change everyday life for the better by ameliorating its supposed constraints, such as taming nature, removing physical drudgery, enhancing enjoyment, adding luxury. As such, the technologies of the toilet have been, and remain, a way to project social status, with the focus on design quality, minimal ornamentation or moving parts, conducive to an historical aesthetics of modernity (cf. Gürel 2008). The main role of technologies here is to hide the messy mechanical control and necessary hydraulic work being conducted, with clean lines that conceal operations and subliminally demonstrate mastery over nature, bringing hygienic orderliness to the world (at least within the confines of the bathroom space). Such designs mean there are also smooth surfaces and fewer visible mechanical elements to harbour germs and disgusting deposits.

Does Touch-free Technology Make a Difference?

> [h]owever natural automated fixtures might seem to engineers, they are all not natural and can even seem alienating to lay users. (Braverman 2010, 15)

A key aim for this chapter was to begin to understand how far digital technology can transform everyday practices of touch. We are concerned to understand how distinct 'smart' technologies, in the form of sensors and software automation, utilises their technicity to transduce the space of shared public toilets differently; how they can make a real difference to how people go to the toilet, and how they feel about toileting activity in shared public spaces. Sensor technologies for touch-free activation are certainly becoming more prevalent in many toilet spaces and are clearly being marketed as powerful tools in modifying the practices of touching. However it is unclear how far touch-free technologies really work in terms of reducing the sense of disgust from direct contact with 'dirty' surfaces shared with strangers, thus making this public space more tolerably habitable.

More conceptually we hope our focus can at least start to provide ways to think about how the technicity of code works in automatically affecting spatiality, for example in the ongoing cultural categorisation of space as 'dirty / clean', 'safe / risky'. Can code itself automate the ordering of the world by ensuring human actors keep 'matter-in-place'? The unacknowledged myth being worked towards is that touch-free sensors and the secondary agency of software can bring into being *fully* automatic space, like shared public toilets that would offer such highly ordered function that surfaces would never become categorised as 'dirty' because 'matter' would never be left 'out of place'. Bathrooms as code/space (cf. Kitchin and Dodge 2011) would thus remake human toileting into a wholly civilised and virtuous practice, preventing it from slipping into an uncivilised or immoral state. Code would provide the ultimate triumph of modernism over nature by completely disconnecting human control over space from the intimate touch of our own corporeality. All embracing software automation also offers up the means to avoid the disgusting animality of others that we are forced to encounter in shared public toilets.

However, in spite of advertising and marketing hype and some potential benefits from touch-free technologies for enhanced convenience and hygiene, their real world implementation is inevitably imperfect. Given that touch-free technologies in shared bathrooms are about enhancing the conventionalised boundaries between 'clean' and 'dirty' in toileting practices by progressively removing the need to touch surfaces, the incomplete and inconsistent way they are deployed means they can only fail in this task. The incomplete deployment of sensors and software across the sequence of activities, including opening and closing doors, means that toileting as a whole can never be rendered fully touch-free and the bathroom fails to become a completely automated code/space. This incompleteness also undermines much, if not all, of the validity of the hygiene discourse used in the marketing of touch-free technologies. If software automation in shared toilet spaces is genuinely about improving cleanliness then comprehensive, 'end-to-end', implementation of touch-free interaction is needed to ensure (near) zero means of germ cross-contamination. Failure at any of the key points in toileting activity by an unavoidable direct touch of a potentially contaminating control surface, such as a door lock, means the complete hygiene chain is broken, that the user's body is no longer safely in the 'clean' category. The results of incomplete and haphazard provision of touch-free technologies in public toilets minimises their value for contaminant control, notwithstanding the fact that in reality some people fail to wash their hands regardless of the technological solutions on offer and normative cultural expectations. Moreover, there is evident inconsistency between touch-free public toilets provision, even within a single institution or the same building, some having no-touch taps and nothing else, others providing only auto-flushing of urinals or hand dryers, and so on.

Touch-free technology is therefore almost always implemented partially, and in inconsistent ways, which can make for user frustration as people are uncertain how bits of an unfamiliar bathroom are meant to work: 'so where do I wave my hands to get some soap?'. The current lack of standardisation of implementation of touch-free sensors can also cause distress for those who struggle with embodied practices in public toilets (Bichard et al. 2008) and can be subtly disabling for some people. Indeed, simpler mechanical bathroom fixtures are better for some users, and the prosaic operation of a tap can be made more problematic with the addition of touch-free technology because the position of the sensor 'eye' is inconsistent across installations, the speed of response and the duration of water flow varies. This may cause mild frustration in a normatively-abled user, but may prevent a physically or cognitively impaired person washing their hands successfully. Another example is how automated air fresheners dispense chemicals that are harmful to some, aggravating asthma symptoms, and in any case merely masking offensive smells to give the impression of hygiene rather than actually purifying the air to remove dust and bacteria.

The partiality of toilet code/spaces is indicative, we would argue, of the modernist hubris that underpins so many 'smart' homes discourses and some of the alluring promise of pervasive computing (Dodge and Kitchin 2009). Such discourses represent a desire for 'tidy space', an excessive orderliness and scientifically rationalised behaviour. This can be read as a 'modern fetish for the *appearance* of hygiene' which

> does not assure the cleanliness it promises. Instead, it merely obscures dirt; indeed, all natural (and finally, historical) processes. Tidiness in fact is only interested in obscuring all traces of history, of process, of past users, of the conditions of manufacture (the high high-gloss). [...] The tidy moment does not recognise process, and so resists deterioration, disease, aging, putrefaction. (Michaels 1990, quoted in Barcan 2005, 9)

The danger then is that toileting is set to become an over-determined activity. It could be argued that attempting to make avowedly simple activities touch-free with digital sensors and software algorithms is simply unnecessary, and an *excess* of automation in the bathroom could be critiqued as an example of disciplining the body through a form of 'technological paternalism' (Spiekermann and Pallas 2006). More tentatively, in step with other discourses extolling the virtues of onrushing 'intelligent environments', bodies should no longer be considered as anonymous entities but instead become identifiable in code in a more differentiated way, with their routine activities available to be recorded. While seemingly far-fetched, assisted living technologies encourage more ambient surveillance technologies deployed throughout the home and the WC is a particular node of concern for certain users, especially the elderly (cf. Dodge and Kitchin 2009). Accordingly, perhaps a few people will actually volunteer to have *sous*veillance built into the toilet bowl, having bathroom sensors and software monitor their every motion, as part of a health-obsessed and bodily performance auditing culture. Yet would most people actually *want* automated, 'intelligent' toilets that identify them and log their 'outputs'? (cf. Braverman 2010). The bathroom and toilet cubicles are one of the few remaining private spaces in modern living, as in many public buildings these are the only blind spots within routine CCTV coverage. Nonetheless they possess the potential to become a new frontier of software surveillance.

More broadly the task of mapping out the places we can touch, the places where we avoid or are compelled to touch, is an interesting challenge for geographers and other social scientists, and we believe our focus on public bathroom spaces and toileting practices is worth exploring further. The arguments presented are only a preliminary consideration of the role of touch-free sensor technologies and software automation to remake the space of toilets as 'clean' code/space by reconfiguring embodied toileting practices. The analysis needs to be extended by drawing upon a wider range of empirics from auditing different shared public toilets, for example within multiple contexts, ages, and levels of usage, and from a qualitatively deeper level of evidence gained by more ethnographic observations of toileting practices and the impacts of technologies on underlying meanings and motivations of performances. Clearly this kind of study of personal practices would require sensitivity given the private nature of toileting and ethical considerations regarding research in shared public space (cf. Barcan 2005, Molotch and Noren 2010).

We believe such studies would be worthwhile to advance understanding of the ways various digital technologies work to mediate direct touch in everyday situations and as such it could contribute to wider understanding in at least four areas of geographical scholarship. Firstly, in terms of affective work looking at emotional and sensual geographies, highlighting how the tactile nature of spatial experiences are changed by sensors. Secondly, it could contribute useful empirical material using ideas around non-representative practices in public environments, particularly in relation to technological control over human bodies and how this is often deflected or sometimes resisted. Using ontogenic notions one could see how toilets come into being as spaces of techno-social practice. Thirdly, such work can advance an understanding of the spatial and social implications of pervasive computing by mapping out how and why the 'automatic production of space' is likely to remain partial, using toilets which are vital but overlooked spaces. The problems of putting code to work in mundane places like public toilets, and the fact that it is so incomplete and inconsistent, actually makes it a fascinating site for doing software studies (cf. Kitchin and Dodge 2011). Lastly, this work speaks directly to the changing the nature of what it means to human. As such it can contribute to debates on post-humanism in which technologies of touch change embodied relationships with the material landscape. Is automation as code/space always going to be imperfect, and will the fetishistic desire for fully touch-free interaction ever be realised? Even if code/spaces built with touch-free sensors and complete software automation were realisable, the question remains whether users would actively *want* them, given the deeper psychological impacts that might result from such corporeal disconnection? Touch-free technologies, therefore, are part of what Robert Macfarlane (2007, 203) laments as the 'retreat from the real.... a prising away of life from place, an abstraction of experience into different kinds of touchlessness'. Software may be able to bring more touch-free spaces into being, but would we ever wish to live a fully touch-less existence?

References

Barcan, R. 2005. Dirty spaces: communication and contamination in men's public Toilets. *Journal of International Women's Studies,* 6(2), 7–23.

Bichard, J., Hanson, J. and Greed, C. 2006. Away from home (public) toilet design: identifying user wants, needs and aspirations, in *Designing Accessible Technology*, edited by P.J. Clarkson, P.M. Langdon and P. Robinson. London: Springer.

Bichard, J., Hanson, J. and Greed, C. 2008. Please wash your hands. *Senses and Society,* 3(1), 79–84.

Braverman, I. 2010. Governing with clean hands: Automated public toilets and sanitary surveillance. *Surveillance and Society,* 8(1), 1–27.

BTA. 2010. *Publicly Available Toilets: Problem Reduction Guide, Third Edition.* The British Toilet Association and Hertfordshire Constabulary Crime Prevention Design Service, <www.britloos.co.uk>.

Cresswell, T. 1996. *In Place/Out of Place: Geography, Ideology, Transgression.* Minneapolis, MN: University of Minnesota Press.

Dixon, D.P. and Straughan, E.R. 2010. Geographies of touch/touched by geography. *Geography Compass,* 4(5), 449–59.

Dodge, M. and Kitchin, R. 2009. Software, objects, and home space. *Environment and Planning A,* 41(6), 1344–65.

Douglas, M. 1966. *Purity and Danger*. London: Routledge.

George, R. 2008. *The Big Necessity*. London: Portobello Books.

Graham, S. 2009 *Disrupted Cities: When Infrastructure Fails*. London: Routledge.

Greed, C. 2006. The role of the public toilet: pathogen transmitter or health facilitator? *Building Services Engineering Research and Technology,* 27(2), 127–39.

Gürel, M.O. 2008. Bathroom as a modern space. *The Journal of Architecture,* 17(3), 215–33.

Hetherington, K. 2003. Spatial textures: place, touch and praesentia. *Environment and Planning A,* 35(11), 1933–44.

Jewitt, S. 2011. Geographies of shit: Spatial and temporal variations in attitudes towards human waste. *Progress in Human Geography,* 35(5), 608–26.

Kitchin, R. and Dodge, M. 2011. *Code/Space: Software and Everyday Life.* Cambridge, MA: MIT Press.

Mackenzie, A. 2006. *Cutting Code: Software and Sociality.* New York: Peter Lang.

Macfarlane, R. 2007. *The Wild Places.* London: Granta.

Miller, W.I. 1998. *The Anatomy of Disgust.* London: Harvard University Press.

Molotch, H. 2008. Peeing in public. *Contexts,* 7(2), 60–63.

Molotch, H. and Noren, L. 2010. *Toilet: Public Restrooms and Politics of Sharing.* New York: New York University Press.

Paterson, M. 2007. *The Senses of Touch: Haptics, Affects and Technologies.* Oxford: Berg.

Spiekermann, S. and Pallas, F. 2006. Technology paternalism – wider implications of ubiquitous computing. *Poiesis & Praxis: International Journal of Ethics of Science and Technology Assessment,* 4(1), 6–18.

Thrift, N. and French, S. 2002. The automatic production of space. *Transactions of the Institute of British Geographers,* 27, 309–35.

Chapter 10

In Close Embrace:
The Space between Two Dancers

Sarah G. Cant

'Let me go where I have not yet arrived.' (Irigaray 1992, 25)

The music is quiet momentarily, and I open my eyes; all that can be heard is the swish of feet moving across the wooden floor, rhythmically with the music. I glance briefly across the space of the milonga [social dance event for Argentine tango]; couples move around the room, absorbed in their dance. Some couples dance closely, whilst others hold each other arms length apart. Silent communication is taking place between each couple; sensed, felt, danced. From a distance an observer might wonder if they are all dancing to the same music, each couple moving in a distinctively individual way around the dance floor, yet they are all dancing tango. I close my eyes again, all the time our dance continues, its improvised form taking us into steps and spaces and sensations, each time different from the last. I am within a tango embrace, connected with my partner, within the music and the dance, yet I will never know precisely where the tango will move me, each time, each moment different to the last. Continuously moving, not arriving, yet always present, always open to what happens.

Argentine tango is an improvised social dance. The couple dance in an embrace, and traditionally there are two roles: the man's and the woman's (Denniston 2007, Savigliano 1995). There are a relatively small number of steps – forwards, back and sideways – a few pivots, turns and spins, and whilst this perhaps over-simplifies what appears to the onlooker to be a complex dance, in essence these moves can be combined in any order to create a new dance every time. Dancing Argentine tango socially challenges both individuals within the dancing couple, 'because of the moment-by-moment improvisational relationship between partners, the tango demands that we pay attention in a way few of us have ever done before' (Fabiano 2002, 1). Each person in the couple must 'listen' to the other intently with their body, in order to communicate (wordlessly) within the embrace and to dance with clear intention and expression. The Argentine tango embrace is shared between two; it provides an opening for 'sense-based spatial experience, intimacy and inter-subjective relations' (Rawes 2007, 49).

Argentine tango is a dance shaped through and by touch, suggested partly by over-lapping etymological origins for the word 'tango', migrating to Argentina from Africa and Europe. These include the African Bantu word 'tang' which means 'to approach, to feel, to touch' (Salmon 1977, 860); a '*tangó*' is a drum of African origin (and dancing rhythmically to *tangó* drumming); and finally, definitions of the Latin terms 'tactum' and 'tangere' include 'to play, handle, touch, feel; to be pleasing; to cause impression, move, excite' (Salmon 1977, 860). The dance also has historically well-rehearsed 'genderings of bodily movement in which, for instance male always leads female' (McCormack 2008, 1826). These two roles, leading and (by implication) following correspond with binary modes of thought in Western philosophy: male/female, active/passive, present/absent, self/other. A feminist position sees these binaries as corresponding to unequal forms of power, status and influence, with the male, active, present, self as dominant across Western culture, language, politics, economics and philosophy (emphasising the masculine construction of the subject), whilst the female, passive, absent, other is silenced or reproduced without her own voice and status, 'defined within and by patriarchy' (Whitford 1991, 16).

Luce Irigaray (1985) seeks to displace these binaries in her search for 'the feminine as defined by and for herself' (Whitford 1991, 16) which she sees as not yet existing within Western philosophy. Rawes (2007, 15-16) summarises Irigaray's perspective on the individual subject: 'although philosophy promotes individual action and experience, it is always identified with male, masculine, of at the very least a 'neutral' subjectivity which does not express the actual physical, mental and spiritual differences of women'. Irigaray (1985, 2004) argues for a different approach to subjectivity, one that recognises the subject through an acknowledgement of the differences between people, 'there might be the possibility of a different, non-masculine discourse' (Whitford 1991, 4). This discourse might emerge through closer attention to the (individual) body, as Braidotti (2003, 44) states, the 'embodiment of the subject is a key term in the feminist struggle for the redefinition of subjectivity. It is to be understood as neither a biological nor a sociological category, but rather as a point of overlap between the physical, the symbolic and the material social conditions'. For Irigaray, sensory perception, and touch in particular 'challenges the importance placed on the autonomous, self-determining and seeing subject in Western thinking. [...] Irigaray proposes a new positive 'economy' of touch and sensation which is created through shared intimate spatial relationships and histories' (Rawes 2007, 48).

Touch is perceived as offering a fluid means of communication in comparison with fixed, dominant (i.e. masculinised) textual modes of expression in language and philosophy. Listening is also important for Irigaray because it implies someone who is respectful of what the other person has to say; if two people listen to each other there is greater scope for communication and recognition of each person on their own basis (without recourse to patriarchal structures). This is at the heart of Irigaray's quest for feminine subjectivities; the recognition of difference between two people, each on their own terms, with neither reducible to a version of the masculine self/same. An emphasis on listening and touch requires individuals to pay attention and 'allow the other to emerge' (Irigaray 1996, 125), there is reciprocity. Irigaray searches for a new 'space of relation' that 'might allow engagements between differences' (Rose 1999, 253), a space 'in between' that 'does not entail solid shape, boundaries, fixity, property and possession, whether of the self or others' (Rose 1999, 254). Although Irigaray initially searches for the spaces 'in between' that would enable the emergence of feminine difference, implicitly the wider project is to 'support the subjectivity of all through active recognition of our interdependence and mutually constitutive activity than to allow the silencing of an other (or group of others) in order to maintain one's own subjectivity' (Lorraine 1999, 21).

Irigaray's quest to go where she has 'not yet arrived' in Western philosophy – to locate differentiated subjects – finds sympathetic bodies within Argentine tango in several ways, including the improvised form of the dance and being open to what might come next without anticipation, and the opportunities the embrace may bring for close communication between two people and for disrupting 'gendered' roles within the dancing couple. This may initially seem surprising when acknowledging tango's historical associations with machismo (Savigliano 1995, Taylor 1998), and popular exoticised, theatrical images of tango dancers that seem to perform gendered choreography very well. However, in practice, much takes place between the dancing couple that 'facilitates ways of moving that do not fit neatly' (McCormack 2008, 1826) into binary categories. Listening and touch shape the dance in many different ways; because the roles in tango *require* each dancer to listen to the other in the embrace, this has the potential to unsettle a binary structure of activity (leading/masculine) and passivity (following/feminine). Irigaray (1996, 121) proposes that 'listening to the other, sparing them some silent time, is respecting his or her breath too'. This chapter is inspired by the writing of Irigaray on sexual difference and subjectivity, touch and listening (1985, 1992, 1996, 2000, 2004). Specifically, it focuses on the embrace in Argentine tango, which many identify as a unique type of embrace within partner dancing (Cara 2009, Denniston 2007, Manning 2007). Although Irigaray emphasises embodied subjectivities, her 'broad-scale cultural analysis' lacks 'physical specificity' (Summers-Bremner 2000, 92); the embrace offers a physical context for exploring embodied subjectivities and the reciprocity of touch between two people dancing.

There are two versions of Argentine tango, one performed for 'export' which is 'flamboyant' and 'danced on stage both locally and internationally to entertain the world' (Cara 2009, 439). The other version is more esoteric, referred to as 'home tango' and 'danced intimately among family and friends or in the numerous *milongas*' (Cara 2009, 439). Cara makes an important point when stating 'home tango does not refer exclusively to tangos performed on home soil. Rather, it speaks of *a way of dancing*' (Cara 2009, 440). Importantly, she notes 'esoteric though they be, the local poetics of home tango are not impermeable. As with all cultural expressions, home tango can be taught and learned across national boundaries' (Cara 2009, 459). This way of dancing, what Cara refers to as 'home tango' or 'esoteric tango', shapes this chapter, which is informed by ethnographic research with dancers and observant participation at *milongas* in Britain, locating Argentine tango within a predominantly non-diasporic context. It provides an alternative cultural position from which to explore the possibilities of moving away from traditional gendered roles, i.e. away from the socially ordered, gendered setting of a traditional *milonga* in Buenos Aires where men lead and women follow (for a discussion of masculinised cultures and the *milonga* in Argentina see Tobin, 1998 and Savigliano 1995, 1998). The cultural position I present here can be interpreted as a 'parallel tradition' (Shay 2008), which refers to the idea that a dance has multiple layers, only one of which is in its 'traditional' location, and other performances of the dance that run in parallel elsewhere are nonetheless authentic.

Writing about dance presents challenges, particularly when an embrace reveals little from the outside, and as Fabiano (2002, 1) suggests within the dancing couple's embrace, 'an encounter takes place between what is inside you and what is inside your partner'. Nonetheless in this chapter I attempt to describe how it feels to dance Argentine tango from an auto-ethnographic perspective. As dancers create tango through embodied, lived experiences, tango accrues new levels of meaning (Borland 2009). To remain at a distance and ignore the interpretation of these meanings, acquired through touch, listening and the intimacy of an embrace, runs contrary to both a feminist engagement with questions of difference, and it would remove the immediacy of touch and sensation, the tactile encounter with another dancer (auto-ethnographic text is depicted in italics).

I begin by looking at social dancing, experiences of touch within dance more widely, and the connections between dancing and Irigaray's perspectives on difference and subjectivity. I then turn to the Argentine tango embrace. Following this I trace a series of embraces with different dancers in the social setting of a *milonga* to consider what it might mean to dance in a close embrace, how the embrace shapes relations between two people, and reflect on aspects of difference and subjectivity.

Social Dancing

Dancing provides an array of sensory, physical and social experiences, from individualistic styles like break-dancing, to group formats such the Cuban *rueda*; from performances of ballet, to partner dances like the Lindy Hop. Bull (1997, 269) highlights some of the physical sensations experienced whilst dancing, 'kinesthetic, visual, tactile and auditory, [...] the shifting of my weight and the changing shapes of my body. I see my surroundings and sense the rush of air past my skin; I hear, and feel, the percussive rhythms of my footfalls.' Thomas (1995, 3) identifies dance as a social activity 'in which many people participate at various stages of their lives, sometimes by themselves, in couples, or in groups, in a range of social settings, from street dancing to dance halls'. Argentine tango in Britain falls within this tradition of popular appeal. For some dancing Argentine tango is a hobby once a week. For others it becomes a way of life, 'daytime' lives are adapted in order to facilitate a deep commitment to classes, practising and attending *milongas*; their pursuit of Argentine tango is likened to a drug, seeking the perfect dance (Goertzen and Azzi 1999, Savigliano 1998).

Dance is a method of cultivating the body, a discipline 'through which it is molded, shaped, transformed and in essence, created' (Foster 1992, 482). In this respect dance is similar to other physical culture pursuits such as sport which Dyck and Archetti (2003, 10) argue are both 'ephemeral creations that are experienced and forgotten or that live on through being reflected upon, witnessed, named, remembered and, quite possibly, repeated.' They emphasise the sociality of dance and sport: 'An individual's embodied discoveries [...] cannot be readily verified or discursively celebrated without the assistance of knowing witnesses'. They go on to suggest that embodied selves cannot 'be easily generated or sustained in sport or dance without the presence and assistance of co-participants' (Dyck and Archetti 2003, 10). There are many dancers in the *milonga*, but only two in a tango embrace, each sustaining their own presence in the dance's ephemeral creation whilst witnessing that of the other dancer.

Dancing, Touching, Two

Irigaray's 'rejection of linearity and [her] embrace of fluidity, its emphasis on developing a new experience of time and space and its use of the body as a base for poetic images' (Kozel 1997, 107) holds great attraction in the context of dance as a cultural form characterised by ephemeral experiences. For Hamera (2001, 231) the reciprocity of touch in the relationship between teacher and student is important in the context of Pilates, which involves 'working with a practitioner in a medium that involves micro-readings and performances of one's body'. Whilst she refers to this as 'a touching which respects the other proffering him/her attentiveness' (Hamera 2001, 231) and the teacher's manipulation of her body into correct Pilates positions involves intimacy; there is a certain 'mastery' in being instructed by an other, when the body has to comply or otherwise be in pain. Summers-Bremmer (2000) explores the relationship between two in a different way, focusing on the duality within the body, 'the two selves the classically trained dancer learns to hold in tension: the self she sees from a distance in the studio mirror – the body as passive instrument, as object of her labours – which would be 'woman' in the traditional binary, and the self she is while dancing, from which the impulse to dance itself arises, equivalent to 'man', the 'active' part of this event' (Summers-Bremmer 2000, 94). To this extent, the duality within the individual demonstrates how dance 'may reflect *and* resist cultural values simultaneously' (Reed 1998, 512). Attention to social dancing is missing in discussions, yet as Manning states, 'there is no touch in the singular. To touch is always to touch something, someone' (Manning 2003, 7).

Dance writers argue that 'dance, perhaps more than any other body-centred endeavour, cultivates a body that initiates as well as responds' (Foster 1995 quoted in Reed 1998, 512). Fraleigh (1987, 59) continues this theme, '[t]he body is a thinking, feeling, acting, expressive whole. We are always expressing something;

at the same time, we are constantly influenced by the otherness of objects and other people as we interact with them'. Many forms of social dancing with a partner, including *salsa* and Western ballroom, appear to reinforce 'restrictive gender roles in society, because the male initiates the relationship and supports his female partner', whereas the woman 'marks time until she can respond to the style and rhythm of the dance established by the male' (Polhemus quoted in Borland 2009, 477). Desmond (1993/94, 37) argues dance produces 'socially constituted and historically specific attitudes toward the body in general, toward specific social groups' usage of the body in particular, and about the relationships among variously marked bodies, as well as social attitudes toward the use of space and time', and she goes on to identify race, gender, ethnicity, nationality and sexuality as 'tenacious categories of identity' mapped onto 'bodily difference'.

For Irigaray, the silencing of the feminine subject is performed through a 'logics of the same', the subject is always masculine 'because of a failure of the symbolic meaning systems of our society to enable female speakers to speak as women or to support them in doing so' (Lorraine 1999, 23). Irigaray's concept of sexual difference would see 'others' (women) speak for themselves and not be disavowed; this is an ethical project which recognises 'our interdependence and mutually constitutive activity' (Lorraine 1999, 21). However, her insistence on two modes of production – masculine and feminine – has been criticised strongly, particularly as it refers to white, western heterosexual constructions of difference (for example see Browne and Rose 2004), although there are possibilities to go beyond her initial conception of the subject and explore an ethics of difference through other relationships and encounters. And Braidotti (2003, 44–45) argues that because Irigaray sees 'sexual difference as the matrix of power does not mean that she neglects or down-plays other differences. On the contrary, [she] broadens the range of her intervention to cover spatio-temporal coordinates and a number of many constitutive relations, including race and ethnicity and especially religion.' Canters and Jantzen (2005) suggest that multiple subjects and other forms of difference, not only sexual, can further develop Irigaray's insights on an ethics of difference. In addition to this, Stone (2006, 1) argues that Irigaray overlooks the 'natural multiplicity within each of our bodies: a multiplicity of forces and capacities such that we are never simply sexually specific'. There are then (at least) two subjects, ways of doing, ways of being.

With reference to the gendering of leading and following, Cara (2009, 454) argues that Argentine tango is 'often misunderstood as a purely 'macho' expression [when it is] a mutual dialogue between equal partners. This contrasts dramatically with historical accounts of the dance' that locate the roles in a binary relationship. Olszewski similarly describes leading and following as 'a conversation between partners, rather than a kinetic fiat issued by the leader' (2008, 70). But a dance has to start, somehow; Manning (2007, 4) states 'tango begins with a music, a rhythm, a melody' and one dancer has to initiate 'a lead, a direction, an opening to which a follower responds'.

'Tango can be discussed and we do talk about it, but like everything it conceals a secret' (Borges in Thompson 2005, 277). For Thompson (2005, 277) the secret is the unseen non-verbal communication that passes between the couple dancing, 'strangers can't imagine how coded intelligence, igniting complex steps, passes from a man to the woman, if they don't talk as they dance'. What Thompson suggests, however is only half of the secret; 'coded intelligence' also passes from woman to man, and the dancing couple could be two women or two men. Briginshaw (2009, 82) explains how Argentine tango 'like many Latin dance styles is imbued with associations of the erotic and exotic and its rendering of heterosexual desire masks the many tensions in its narrative', including the homosexual and homo-social. If Argentine tango, whilst employing the two roles of leading and following, is open to being read as 'transgressive in terms of gender' (Tobin 1998, 98), then 'the opposition between activity and passivity no longer has any meaning' (Irigaray 1996, 126). Irigaray continues: 'communication *between* and reciprocity, as well as respect for one's own gender (never simply one's own since it is engendered and remains partially exterior to one's self), respect for the gender of the other, for listening and silence, require *touching upon* without reduction or seduction, the safeguard of the sensible...' (1996, 126). Touch can reveal much about an individual's personality, feelings and state of mind (Autton 1989), but this, combined with Argentine tango's gendered histories, suggests listening to the other may not be quite so straightforward as Irigaray implies.

Embracing

To embrace someone is to take them in your arms and hold them closely, an embrace is reciprocal. In everyday circumstances, an embrace is an expression of affection, reserved for and shared between those we know well: lovers, friends, family. An embrace can be a deeply sensuous tactile experience, an emotionally moving and intimate encounter between two people. In the context of Argentine tango, an embrace is much more than a quick hug in greeting or departing, it is sustained for the period of at least one tango dance, often a '*tanda*' (a set of three or four tangos lasting up to twelve minutes, usually danced with the same partner). An embrace is a direct, tactile encounter with another person, that potentially 'has the capacity to dissolve boundaries, to make proximate that which was far away, and in doing so not only rearrange our metaphysics of intimacy and distance, but post a danger to any and all systems of order that rely upon distinction and separation' (Dixon and Straughan 2010, 454). It may become possible to identify 'questions of otherness, strangeness, difference and the emotions moving within' (Krappala 1999, 33) an embrace.

The Argentine tango close embrace has been described as 'heart to heart', likening the positioning of the two bodies embracing as they dance together to that of a 'natural, loving hug' where 'two people generally stand directly in front of each other, with their shoulders parallel and the centres of their bodies aligned'

(Denniston 2007, 31). This embrace requires each dancer to give to the other 'their complete and undivided attention' (Denniston 2007, 31), upper torsos touch and arms encircle to complete the dancers' embrace (Figure 10.1). Denniston argues this is different to holds used in other forms of partner dance (including Western ballroom) as the embrace is 'the most intimate, personal and perhaps even emotionally challenging position in which to dance' (2007, 31). This intimacy is what makes Argentine tango different to other forms of dance. For some it is why they choose to dance tango, as Babitz (1992, 129) suggests: 'for someone like me, who'd been fleeing from intimacy all my life, [I was] being drawn to the tango because it was *about* intimacy – the art of it'.

Figure 10.1   'Dancing in close embrace'
Source: author photograph

Whilst different forms of social dance claim intimacy, for example West Coast Swing has been described as 'an intimate conversation between two people' (Callahan 2005, 3), often 'intimacy' refers to communication between leader and follower, not the additional physical intimacy of a close embrace. Studio *salsa* provides another example, where physical points of connection – principally pressure between hands – 'remains a discreet point of contact, as opposed to the bodily embrace employed in tango' (Borland 2009, 490). The Argentine tango close embrace requires a specific kind of physical commitment between two bodies that is more intimate than other dances, before any other type of connection is made between the two people dancing. This physical intimacy, heart to heart, offers further points of sensuous, emotional and affective connection for the dancers across a range of registers.

Figure 10.2 'Open embrace'
Source: author photograph

It is also possible to dance Argentine tango in an 'open embrace', with a degree of physical space between the torsos, arms extended to hold each other, forming a more visible circular shape to the embrace (Figure 10.2). However, even though torsos may not be touching, primary communication within an open embrace is also 'from the centre' or 'the heart' (Denniston 2007), or in approximate anatomical terms, the solar plexus. In an open embrace, whilst there is not the direct physical intimacy of torsos, there remains intimacy through communication between the two dancers. The arms in an open or close embrace should not push, grip or be rigid; hands do not push or pull. Dancers communicate through the projection and sharing of energy from their centres, a sense of touch which travels through the body, through the arms, through the legs. A lead invites a response from the other dancer; in skilled and communicative dancing, the slightest movement in a leader's centre can invite movement in a specific direction for the follower's leg. The hands and arms do not instruct, and this makes tango different to many other partner dances, such as ballroom, where hand pressure on the back communicates intentions and directions.

The *milonga* brings together people from many different backgrounds, requiring strangers to embrace and set aside their everyday inhibitions. One of the first decisions to make when starting to dance with a new partner at a *milonga* is how close the embrace will be. To dance socially in an embrace is to experience tactile encounters as a couple who may not be lovers, friends or family. To dance in an embrace is to foster a particular kind of intimacy that seemingly oversteps the 'norms' of encounter with strangers. In other aspects of everyday life one is often careful to avoid 'invading someone's privacy (in public places)' (Laurier and Philo 2006, 357) as this could be perceived as disrespectful. This sense of maintaining a distance between strangers was exemplified in an interview with a male dancer:

> There is a certain amount of intimacy that you would just not do in normal life. You have friends or people that you know to a certain degree. But equally there are other dancers that you don't know, and for ten to fifteen minutes dancing you embrace in a way that's not something that you would otherwise do. For example if you are underground in the tube you stay away trying not to touch anybody at all! (April 2010)

In the beginning close embrace felt scary with people he didn't know, but he emphasised 'after a while you just get used to it and, it feels normal and natural' (Interview, April 2010). The close embrace is typically confused by non-dancers with eroticism and sex. However, 'while certainly sensual and seductive, the home [social] tango does not privilege sexuality. Intimacy is favoured over sexiness' (Cara 2009, 454). This point was emphasised in an interview with a female dancer:

> There's something really lovely about being held by somebody who is actually holding you in an incredibly comfortable way. And you're both moving together in a completely non-sexual way because it's about music and dancing and kind of sociability. But it's also really sensual. (March 2010)

Fabiano elaborates further, 'most of the time the intense feeling is coming from the sensation of the tango connection itself, or from their love of the dance itself, not from sexual attraction' (2008, 3). Images of tango as an erotic dance are constructed through performances of 'export tango' (Cara 2009), not the social setting of the *milonga*.

The *Milonga*

Embraces take place across generation divides and differences of ethnicity and class, short dances with tall, large dances with small. As a social event, dancers come to the *milonga* with the intention of dancing with different people over the course of the evening (and people will have their personal preferences for dance partners). In many *milongas*, between each *tanda*, a very short piece of music is played, the '*cortina*', a signal to leave the dance floor and change partners (or rest). With our chosen dance partner, at the start of a *tanda* an unspoken negotiation takes place, how close shall we dance? Upper bodies adjust to become close, or to maintain a distance between two people who may or may not know each other. Often this negotiation is smooth and bodies fit together seamlessly at whatever the chosen distance, sometimes it can become fraught with tensions as dancers become aware of their differences. To dance tango requires a certain confidence that you are potentially able to dance with anyone in the *milonga*, that through your love of tango, you are at ease with others. We are here to dance, I am happy to hold you, even if we have never spoken or danced together previously. However to hold someone, perhaps someone you have not met before, requiring intimacy, there is vulnerability.

The commencement of a dance is a powerful moment, as Thompson (2005, 276) explains, dancers 'pause as they start. ... They take each other carefully in each other's arms and pose for a few moments on the dance floor, as if in deep thought. Only then do they move and become one with the music. ... [Letting] a few measures go by [...] to hear the music and enter into the climate of its theme'. In these first moments of the embrace we meet the other dancer directly, our knowledge is physical, of torsos, arms, hands. We are aware of differences, of gender, age, height, width, smell, sweat, and more. In the dance some of these differences may dissolve, others remain, and new ones appear. We do not know what shape the dance will take, but hope that it will be connected, in touch. By connection, I am referring initially to the physical sensations that enable each to read the other's body and the signals exchanged between upper torsos, the lead and follow effect within the dance. But, as Taylor suggests, even when holding each other closely, connection is more than physical she contends:

> Tango bodies are thrown [... into an] embrace that paradoxically can establish the most intimate of links or none at all. It can divorce not only the bodies but also the parts of these bodies from each other. This is a performance of rupture in the same way that it could be – and sometimes is – a performance of tenderness. (Taylor 1998, 65)

The embrace combines the sensual and the emotional, tango folding, touching within, between and inside the spaces of the dance. In the embrace we meet the other, and we are aware of their wonder, particularly in close embrace. Let us dance.[1]

The First *Tanda*: our First Dance

> *We haven't danced together before. The leader holds out their left hand, which my right hand clasps firmly but softly, and then without hesitation we fit into a close embrace, we lean in a little, the centre of ribs connecting, my left arm around their right shoulder, their right arm holding me closely. Together we shift our weight from left to right, and back again, and again. We take a side-step, then start to walk. Then they lead an ocho [a figure of eight step] and but there's pushing and pulling on my back by their right arm. This is uncomfortable (and incorrect). Despite our bodies being held closely together, we are not connected. I can feel tension travelling down the leader's left arm into my hand and this hold becomes a grip. However much I flex my fingers, try to move back and suggest a softer hold, it is unnoticed. I am now in a vice. This is not a good dance. The tango ends, we step apart and smile awkwardly at each other; as the next tune begins I decide to open the embrace and keep this partner at a distance. The next dance is better, but I have to work really hard – the leader seems to keep trying to pull me out of my axis, my balance. When they ask, somewhat dejectedly 'and a third?' If I'm honest, I want it to end, but I'm polite and smile and say yes – I don't want to hurt their feelings. We chat afterwards; it turns out this leader has only had a handful of lessons and hasn't yet been taught to dance in close embrace.*

Cortina: Intimacy, Touch, Space and the Unfamiliar

I have chosen to follow this evening, but am acutely aware of the gendered imposition this might place on my role. There is a risk of conforming to expectations, fitting in with historical binary perspectives on social life. However, in my wider experience of following, this is not a passive docile role but an active half of the dance, committing energy and presence in connection through interpretation, suggestion and embellishment. But in this encounter we struggled. It began too intimately; only hindsight offers that perspective. The unfamiliar to them was dancing with the other, possibly even touching the other. Despite each dance being different, it did not feel new, it was a wonder-less experience resounding in isolation. Battles of strength, power and dominance – familiar themes – the embrace became a space in which gendered roles were rehearsed, but also reversed, enacted and exerted by each person within the embrace. This dance was an 'either-or', the embrace

---

1   This *milonga* is a composite of various venues and dances.

revealing a binary and communication was not fluid, not reciprocal. The roles of leading and following were defined by what they were not (the other). They were in competition, not two singularities respectful of the others difference. Dancing Argentine tango socially requires individuals to set aside habits of non-interaction with strangers, but critically within the embrace there should also be respect for individual integrity, which (perhaps) paradoxically can lead to a stronger sense of being there for the other person in the dance when it is combined with listening. But dancing with strangers can be difficult, the unfamiliar adding to the stress of the encounter. Fear of the unexpected can cause tension, and the more a body tries to relax, the more it contracts through concentration and effort to be correct and communicating with another through touch becomes rigid, unsatisfactory. We did not listen to each other.

Second *Tanda*: a Playful Dance

The music is exciting, there is lots of space in the middle of the room, and we have such fun dancing in open embrace with big movements: luxurious long ochos, playful ganchos where legs hook and interlock, and boleos, my leg swinging high into the air. The grins and laughter are spontaneous; we know that we feel the music in a similar way and there is plenty of space for embellishment within our dance; a cheeky heel tap here, a hip movement there. We are dancing at a distance, in a wide, lightly held embrace, but are so connected. There is nothing I would want to be different in this dance, and lead and follow start to merge. I am acutely aware of the slightest movement the other makes, and vice versa, and together it all feels right.

Cortina: Wonder, Trust, Space

In this encounter we danced playfully. Nachmanovitch (1990, 43) suggests 'play is the free spirit of exploration, doing and being for its own pure joy'. Tango provides a discipline for playing, 'of mutual awareness, consideration, listening, willingness to be subtle. Trusting someone else can involve gigantic risks, and it leads to the even more challenging task of learning to trust yourself' (Nachmanovitch 1990, 97). The lead, the sensation that makes my leg fly into the air, causes that effect because I trust my partner. There is no tension in my body, I relax, I know that this leader is aware of the space available on the dance floor. We will not crash into other dancers. Not all leaders have that trust. Although we were dancing in an open embrace, this was an intimate experience. The trust enabled by this embrace was facilitated by two dancers open to dialogue through touch. 'I am listening to you: I encourage something unexpected to emerge, some becoming, some growth, some new dawn perhaps. I am listening to you prepares the way for the not-yet-coded, for silence, for a space for existence, initiative, free intentionality, and support for your becoming' (Irigaray 1996, 117). This embrace created a space through which we recognised the other on our own terms, through good communication. We listened.

Third *Tanda*: a Heart-less Dance

> *As we dance, although we are connected physically and to all other intents and purposes the dance seems to be going well, there is something missing, something being drawn away, drawn back. I feel a negative, empty energy projected from this partner towards me. We are dancing in close embrace, but there is distance between us. I'm still following, but as the dance progresses my mood shifts down, and down. How is it possible that fabulous music is playing and everyone else in the milonga seems to be dancing beautifully together whilst I'm experiencing one of the strangest dances I've ever had? It isn't a physically terrible dance, but one where the heart is not there and I'm wondering if they are aware of this projection onto me. But I can't ask, we don't talk whilst dancing. It isn't an uncomfortable embrace that makes me want to stop dancing, but it's a puzzle. I try to give more energy to the dance but this person is like a shock absorber, flattening, seemingly dissipating anything positive that might touch them. We are dancing together, but we are dancing in solitude. So many things go unspoken. At the end of the three dances we stop and move away from the dance floor, to different parts of the room. I look round for someone else to dance with, someone I know, someone I hope can restore my feeling for tango.*

Cortina: Reflection on Separateness and Distance in Close Embrace

This dance was lost; somewhere the feeling for the dance disappeared. Tango is danced wordlessly, through touch we listen to the other's body, but on this occasion there was silence, barriers, an emotional fortress surrounding the other. In this silence, can we search for what touches the other? Krappola (1999, 10) suggests that Irigaray would contend that it is the wordlessness that makes visible the discursive spaces between 'this body and that of the other.' But there are different kinds of wordlessness. Julie Taylor recounts a tango class taken in Buenos Aires:

> "Where" the *maestra* asked, "do you look for the sensual? You can look for it, you can dance it in very different ways. Through tenderness, through energy, through anger [...]". Categories dilated again. They overlapped.

> We had come to take each other into account in an intricate way, a way that involved our bodies. The physical had been, after all, our initial mode of knowing each other. [...]

Dancing in the arms of each other, we each regained bodies we had lost, some of us in our histories, others in the days at work or in families, before we stepped onto the dance floor. (Taylor 1998,106–7)

So perhaps when a dance is not connected, not felt in the heart, it is because of a wonder-less embrace, that this time reveals a dancer as distant, separate, and whilst defined through their interaction with the other, not recognising the other on any level. This was not an intimate embrace, although we held each other closely. In this dance I followed, but with no opportunity to do otherwise, I was unable to assert my presence and be heard on my own terms. Gendered constructions of leading and following accompanied this dance, there was no acknowledgement of difference through listening.

Fourth *Tanda*: a Wonder-full Dance, Leading to a Fifth *Tanda*

It's almost the end of the evening; as this leader comes over to me, they apologise profusely as they ask: "I'd love to dance with you, but my t-shirt... I'm sorry, would you? Do you mind?" The t-shirt is sticking to their torso, a dark patch at the bottom of their ribs, another on their back. The signs of having danced all night in close embrace; damp marks where torso and hands have been in contact. This is someone I've wanted to dance with all evening; if I refuse it may be a long while before I see them again. I know that this is someone I want to dance in close embrace with, despite the sweat... Immediately there is a strong connection; I close my eyes and we start to dance. I do not sense the sweat. I feel so safe and relaxed in their arms, I trust this lead immediately; they pick up on this feeling quickly and our dance simply flows, as I interpret their clear lead and all the things that are supposed to work do, when you are dancing well. It is a wonderful dance; a wordless, tactile dialogue that enables each dancer to communicate clearly and recognise the other on their own terms. At the end of the third dance the leader looks at me and I simply nod, and we carry on dancing, oblivious to sweat; three more dances flow. Each small movement, shift of weight, is beautiful, subtle and sensitive, each large movement is fluid and soft. This is amazing. And then we have to stop because the music has stopped. The evening is over.

The Lights Come Up at the End of the *Milonga*

This last embrace, this last dance, has been almost everything that a tango could be. In our ease of dancing together was the recognition of two singularities (Grosz 2005) that, within the space-time of the dance enabled the other liberty, whilst simultaneously accepting losing ourselves by giving ourselves (Irigaray 2004) to the embrace. The close embrace created, metaphorically, an open space where, through the improvised nature of the dance, our attention to 'sensation, feeling, impulse in the process' (Tufnell and Crickmay 2004, 289) was heightened. There were two roles, but no dominance or violence towards the other; positive energies met in the middle of this embrace. As a follower this was an empowering dance. I was aware of being heard and I listened easily to the leader. It was an embrace in which I felt completely comfortable, able to dance as myself, not pushed or contained within a space where I did not want to be.

Embracing Argentine Tango

Dancing these *tandas* revealed some hidden aspects of embraces and what emerges through the dance, how emotions can be communicated and shared through the tactile nature of Argentine tango, holding someone in your arms and dancing. Two roles bring layers of feeling to the dance, leading and following take on different complexions at different moments within *tandas*. The embrace – close or open – is challenging because it 'involves a great deal of concentration, creativity and precision of each dancer' (Park 2004, 13). If one dancer metaphorically steps away from the embrace whilst dancing and stops listening, the connection breaks down. In successful embraces the dance is a conversation, a listening to the other, and paradoxically the two roles overlap whilst enabling dancers to express their subjectivity and be acknowledged. Fabiano (2010, 1) suggests 'a successful tango connection, regardless of style or musical interpretation, is one in which information flows back and forth, replacing the perception of two with the awareness of one'. I interpret this one as the embrace, a space of meaningful encounter that when it works well is not loaded with gendered expectation or restriction. This is not the 'masculine one', the subject of the binary, but a third element within the dancing couple. It is, through the embrace, a 'space in between'. There are qualities of leading and following that enable conversation between the two to flow better – as the *tandas* danced above imply – some dancers are more skilled at recognising and working with differences, whichever role they choose to dance, to reach a state of recognition of the self and of the other, each on their own terms.

Each tango embrace within the social dance has the potential to offer a 'surprise of the unexpected encounter, the productivity of the encounter that welcomes the future openly' (Grosz 2005, 166). Dancing as a follower with male and female leaders, of different ages, heights and build, the mapping of bodily difference and subjectivity in each *tanda* was actually framed around dancing ability and skill, proficiency and sociability as a dancer, and willingness to make a connection with another person, not other markers of social difference. The decision to dance in a close or open embrace was not an indicator of intimacy, this could be achieved in an open embrace as equally as distance between two could be present in a close embrace. The tension in the background of the dance – the historical gendered binary – is manipulated by dancers who recognise the embrace has the potential to facilitate presence and absence for both leader and follower. Whilst I have focused on the experience of following, the tangos danced in this chapter have enabled attention to tango roles without seeking to essentialise gendered relations, through the exploration of what emerges within each embrace. In 'good' dances, leading and following are not opposites, they are not equal (which would imply the masculine same), but they are different uses of energy that are shared, communicated. There are pure moments in tango when two bodies, energies and their associated spatialities within the embrace connect completely and effortlessly in the dance. Everything becomes fluid and light, and perhaps paradoxically, very present: we are here, dancing together in each other's arms. Such moments are special because they are not planned, they emerge spontaneously through the improvised form of the dance, dancing intuitively but dancing in a way that recognises leader and follower are listening to each other.

Irigaray (1996) points out, however that listening does not mean comprehension, which would imply complete, controlling knowledge of the other, and would no longer require people to pay attention to each other. Listening to each other's body in the embrace, within tango's relational infrastructure, the dancer is attentive:

> I am attempting to understand and hear your intention. Which does not mean: I comprehend you, I know you, so I do not need to listen to you and I can even plan a future for you. No, I am listening to you as someone and something I do not know yet, on the basis of a freedom and an openness put aside for this moment (Irigaray 1996, 116–117).

To dance Argentine tango is to be acutely aware of the present moment, without anticipating what might come next; anticipation restricts what might happen and closes down communication between two. Much of what Irigaray writes is about communication and the ways in which this might be facilitated between two people. In a wordless embrace, dancing Argentine tango, communication between two (should) 'respect the other proffering him/her attentiveness. [...] This *touching upon* asks for silence. To allow the other to emerge, silence is necessary' (Irigaray 1996, 124). Whilst Argentine tango is caricatured as a dance of seduction, in practice to dance well and be connected physically and emotionally with your partner in a respectful manner is to respond to the touch of the other within the embrace, regarded not as 'appropriation, capture, seduction – to me, toward me, in me – nor envelopment. Rather it is to be the other's awakening to him/her and a call to co-exist, to act together and dialogue' (Irigaray 1996, 125).

On reviewing the *milonga*, favourite embraces were experienced as such because a partner felt good to dance with. And afterwards feeling rewarded, recognised and different, in a good way; one which enabled the dancer to feel complete and 'whole' from the experience of a physically and emotionally connected dance. The Argentine tango embrace offers but a small inroad into Irigaray's concepts of difference and subjectivity. The language of tango, Fabiano (2002, 2) argues, requires us 'to maintain focus through our heightened senses of touch and hearing – this sensory focus keeps us "in the moment" with our partners'. The embrace, through which the dances' sensual richness emerges, 'makes a thing a tango and not something else; that which, if it were missing, would no longer make a tango a tango' (Gobello 1980, quoted in Cara 2009, 438). The embrace offers a 'fleeting touch of what has not yet found a setting', and requires dancers to be 'always open to what happens' (Irigaray 2004, 155). The embrace is always a temporary space, a set of fleeting experiences, but one that is always shaped by the two people in its hold and the multiple ways in which each dancer acknowledges the presence of the other (or not). In a comfortable embrace, when both dancers respect each other's contribution to the tango and share a commitment to communicating clearly and sensitively, there is potential for the embrace to become a space 'in between' wherein two people 'co-exist, to act together and dialogue' (Irigaray 1996, 125). The Argentine tango embrace touches each individual, it is never a neutral experience, but may enable the recognition of singularity, the difference of the other.

References

Autton, N. 1989. *Touch: An Exploration.* London: Darton, Longman and Todd.

Babitz, E. 1992. Bodies and souls, in *Sex, Death and God in L.A.*, edited by D. Reid, Berkeley, CA: University of California Press, 108–50.

Borland, K. 2009. Embracing difference: Salsa fever in New Jersey. *Journal of American Folklore,* 122(486), 466–92.

Braidotti, R. 2003. Becoming woman: or sexual difference revisited. *Theory, Culture & Society*, 20(3), 43–64.

Briginshaw, V.A. 2009. *Dance, Space and Subjectivity*. Basingstoke, England: Palgrave Macmillan.

Browne, K. and Rose, G. 2004. An exchange of letters provoked by Luce Irigaray's *I Love to You* in *Geography and Gender Reconsidered*, edited by the Women and Geography Study Group (RGS-IBG), CD-ROM, pp. 156–68.

Bull, C.J.C. 1997. Sense, meaning and Perception in three Dance Cultures, in *Meaning in Motion: New Cultural Studies of Dance*, edited by J. Desmond. Durham, NC: Duke University Press, 269–87.

Callahan, J.L. 2005. 'Speaking a secret language': West Coast Swing as a community of practice of informal and incidental learners. *Research in Dance Education*, 6(1/2), 3–23.

Canters, H. and Jantzen, G.M. 2005. *Forever Fluid: A Reading of Luce Irigaray's 'Elemental Passions'*. Manchester: Manchester University Press.

Cara, A.C. 2009. Entangled tangos: Passionate displays, intimate dialogues. *Journal of American Folklore*, 122(486), 438–65.

Denniston, C. 2007. *The Meaning of Tango: The Story of the Argentinean Dance.* London: Portico Books.

Desmond, J.C. 1993/1994. Embodying difference: Issues in dance and cultural studies. *Cultural Critique*, 26, 33–63.

Dixon, D.P. and Straughan, E.R. 2010. Geographies of touch/touched by geography. *Geography Compass*, 4/5, 449–59.

Dyck, N. and Archetti, E.P. 2003. *Sport, Dance and Embodied Identities*. Oxford: Berg.

Fabiano, S. 2002. *The Essential Tango: Dancing in the Moment*, Tango Mercurio Online Resources, http://tangomercurio.org/ar-essential.html [accessed: 2 March 2011].

Fabiano, S. 2008. *Passion Container*, Tango Mercurio Online Resources, http://tangomercurio.org/ar-passion.html [accessed: 2 March 2011].

Fabiano, S. 2010. Tango: A deeper look. *International Journal of Healing and Caring*, 10(2), 1–5.

Foster, S.L. 1992. Dancing bodies, in *Incorporations*, edited by J. Crary and S. Kwinter. New York: Urzone and MIT Press, 480–95.

Fraleigh, S.H. 1987. *Dance and the Lived Body: A Descriptive Aesthetics*. Pittsburgh, PA: University of Pittsburgh Press.

Goertzen, C. and Azzi, M.S. 1999. Globalization and the Tango. *Yearbook for Traditional Music*, 31, 67–76.

Grosz, E. 2005. *Time Travels: Feminism, Nature, Power*. Durham, NC: Duke University Press.

Hamera, J. 2001. I dance to you: reflections on Irigaray's *I Love to You* in pilates and virtuosity. *Cultural Studies*, 15(2), 229–40.

Hamera, J. 2007. *Dancing Communities: Performance, Difference and Connection in the Global City*. Basingstoke, England: Palgrave Macmillan.

Irigaray, L. 1981. And the one doesn't stir without the other. *Signs*, 7(1) 60–67.

Irigaray, L. 1985. *Speculum of The Other Woman* (translated by G. C. Gill). Ithaca, NY: Cornell University Press.

Irigaray, L. 1992. *Elemental Passions* (translated by J. Collie and J. Still). London: Routledge.

Irigaray, L. 1996. *I Love to You: Sketch of a Possible Felicity in History* (translated by A. Martin). London: Routledge.

Irigaray, L. 2000. *To Be Two* (translated by M. H. Rhodes and M. F. Cocit-Monoc). London: Athlone Press.

Irigaray, L. 2004. *Ethics of Sexual Difference* (translated by C. Burke and G.C. Gill). London: Continuum.

Kozel, S. 1997. 'The story is told as a history of the body': strategies of mimesis in the work of Irigaray and Bausch, in *Meaning in Motion: New Cultural Studies of Dance*, edited by J. Desmond. Durham, NC: Duke University Press, 101–9.

Krappala, M. 1999. *Burning (of) Ethics of the Passions: Contemporary Art as a Process*. Helsinki: University of Art and Design Helsinki.

Laurier, E. and Philo, C. 2006. Possible geographies: a passing encounter in a café. *Area*, 38(4), 353–63.

Lorraine, T. 1999. *Irigaray and Deleuze: Experiments in Visceral Philosophy*. Ithaca, NY: Cornell University Press.

Manning, E. 2003. Negotiating influence: Argentine tango and a politics of touch. *Borderlands e-journal*, 2(1), 10, <www.borderlands.net.au/issues/vol2no1. html>.

Manning, E. 2007. *Politics of Touch: Sense, Movement, Sovereignty*. Minneapolis, MN: University of Minnesota Press.

McCormack, D.P. 2008. Geographies for moving bodies: Thinking, dancing, spaces. *Geography Compass*, 2/6, 1822–36.

Nachmanovitch, S. 1990. *Free Play: Improvisation in Life and Art*. New York: Jeremy P. Tarcher/Putnam.

Olszewski, B. 2008. *El Cuerpo del Baile:* The kinetic and social fundaments of tango. *Body & Society*, 14(2), 63–81.

Park, C. 2004. *Tango Zen: Walking Dance Meditation*. Maryland: Tango Zen House.

Rawes, P. 2007. *Irigaray for Architects*. Abingdon, England: Routledge.

Reed, S.A. 1998. The politics and poetics of dance. *Annual Review of Anthropology*, 27, 503–32.

Rose, G. 1999. Performing space, in *Human Geography Today*, edited by D. Massey, J. Allen and P. Sarre. Cambridge: Polity Press, 247–59.

Salmon, R.O. 1977. The tango: Its origins and meaning. *Journal of Popular Culture*, 10(4), 859–66.

Savigliano, M.E. 1995. *Tango and the Political Economy of Passion*. Boulder, CO: Westview Press.

Savigliano, M.E. 1998. From wallflowers to femme fatales: Tango and the performance of passionate femininity, in *The Passion of Music and Dance: Body, Gender and Sexuality*, edited by W. Washabaugh. Oxford: Berg, 103–110.

Shay, A. 2008. *Dancing Across Borders: The American Fascination with Exotic Dance Forms*. Jefferson, NC: McFarland & Company.

Stone, A. 2006. *Luce Irigaray and the Philosophy of Sexual Difference*. Cambridge: Cambridge University Press.

Summers-Bremmer, E. 2000. Reading Irigaray, dancing. *Hypatia*, 15(1), 90–124.

Taylor, J. 1998. *Paper Tangos*. Durham, NC: Duke University Press.

Thomas, H. 1995. *Dance, Modernity and Culture: Explorations in the Sociology of Dance*. London: Routledge.

Thompson, R.F. 2005. *Tango: The Art History of Love*. New York: Vintage Books.

Tobin, J. 1998. Tango and the scandal of homosocial Desire, in *The Passion of Music and Dance: Body, Gender and Sexuality*, edited by W. Washabaugh, Oxford: Berg, 79–102.

Tufnell, M. and Crickmay, C. 2004. *A Widening Field: Journeys in Body and Imagination*. Alton, England: Dance Books.

Whitford, M., 1991. *The Irigaray Reader*. Oxford: Blackwell.

Chapter 11

Intra-body Touching and the Over-life Sized Paintings of Jenny Saville

Rachel Colls

Introduction

Recent work in geography on touch and touching has drawn attention to what Dixon and Straughan (2010, 450) have described as the 'myriad interrelations that are thought to exist between and among the 'interiority' of the human body – that which we may refer to as the psyche or even the soul, but also the meat, flesh and bones of the soma – and an 'exterior' world of other people, life forms and objects'. This work includes that which considers how bodies touch and are touched by different 'others' (broadly understood as objects, human and non-human bodies, materials and 'the environment'). For example, this form of touching relation is demonstrated in Obrador-Pons' (2007) work on bodies being touched by the sun and sand on nudist beaches, Ingold's work (2004) on walking and the feet's relation with the ground, Macpherson's (2009) account of touch with reference to people with visual impairments and hill walking and Lea's (2009) work on bodies touching bodies in massage therapy. It also includes work in emotional geographies that focuses on the interrelations between the interior and exterior of bodies by focusing on how bodies 'feel', 'touch' and are touched by other bodies and spaces (see Davidson, et al 2005).

However, what is often overlooked in such work is any indication that bodies touch themselves or that bodies touch according to the sized or sexed specificities of their bodies. From these previous examples consider hands smoothing sunburnt breasts, thighs and toes rubbing together when walking and rubbing tired eye lids after a long day in the field. This chapter, therefore, will develop an explicitly feminist account of touch by focusing on how bodies touch themselves, or what I have termed 'intra-body touching'. Such an understanding of touch considers the intra-corporeal relations that occur upon, across and within a body. This has resonance with recent work on 'the visceral', described by Hayes-Conroy and Martin (2010, 272) as 'the sites of the human body in/through which feelings, sensations, moods, states and so on are experienced', which has focused on the relations between the biological and the sociological and the socio-political significance and affective experience of touching and being touched by food when it comes into contact with 'the body' (see also Longhurst et al 2009). However, it is noticeable in this work that the relationships between the interior and exterior spaces of the body are still mediated through the involvement of an 'other' i.e. food. Intra-body touching, therefore, focuses attention upon specific geographies and relations of touch whereby spaces are created upon and within the body when bodies touch themselves.

In this chapter, the nature and significance of intra-body touching will be demonstrated through an interrogation of two over-life-sized paintings (*Branded* and *Propped*; see Figures 11.1 and 11.2) by the artist Jenny Saville. Her paintings present the topographies of a female fleshy body through detailed observations of bodily surfaces and orifices which include breasts hanging, hands grabbing and fat rolling and pressing upon itself. These relations are not only empirically significant i.e. they draw our attention to the sexed and sized specificities of the matter of bodies (see Colls 2007); they are also useful intellectually for making sense of how bodies, things and matter relate 'with' and 'to' each other.

In order to theorise the significance of intra-body touching in Jenny Saville's over-life sized paintings, Luce Irigaray's (2004) notion of the 'morpho-logic' is utilised. The 'morpho-logic' is a language that is based upon the morphology of the female body and is deployed by Irigaray in recognition of its denial in phallocentric philosophical thought. Identifying the phallocentrism of particular knowledges, in Irigaray's case a range of Western philosophical texts (including the work of Merleau-Ponty to be discussed in this chapter), involves acknowledging the power and privileging of 'Man' or a 'male world view' that situates 'Woman' and 'the feminine' in a subordinate or 'othered' position. It also means highlighting what is called the phallogocentrism of such knowledges which involves recognising the denial, effacement and/or containment of a specific female/feminine imaginary (see Whitford 1991). In this case the 'phallus', whether literally or figuratively, acts as the standard against which all else is measured, i.e. woman exists as a 'lack' or an 'absence'. Irigaray (1991, 96) argues that this contributes to the production of a specific 'morpho-logic' that involves 'the imposition of formations which correspond to the requirements or desires of one sex as the norms of discourse and, in more general terms, language (*langue*)'.

In response to this, Irigaray develops a 'feminine imaginary' or new kind of 'morpho-logic' through the use of the *mucous*. The mucous, exemplified by the lips of the labia, illustrates the ways that women's bodies are always and already 'touching' and in places and contexts that may not always be visible. In using such a strategy, she is not advocating the need for an anatomically deterministic account of sexed subjectivity. Instead, by revealing the phallocentric nature of knowledge production she is, in Robinson's (2006, 58) words, 'breaking the tautology where only men's sex is spoken'. Moreover, the morpho-logic of the mucous also provides us access to touching relations that blur the boundary between self and other (see Young 1990). The significance of this 'blurring' can be found in Margrit Shildrick's (2001) work on co-joined twins (described by her as a 'non-normative morphology'). She develops an ethical account of touch in order to problematise the ways that the 'specular normative economy' privileges the distinction and separation of the subject by what is visible. Shildrick (2001, 392) further suggests that 'the corporeal ambiguity of touch that disrupts the distinction between self and other' enables us to 'come face to face with the leaks and flows at the boundaries of and the vulnerabilities within, our own embodied being'.

For 'fat' bodies, another non-normative morphology that produces touching relations that are not necessarily visible, this ambiguity offers an alternative relation with 'fat' that is not dependent on estrangement and distancing (i.e. fat as the 'other'). Instead, relations of touch premised upon 'proximity' and 'intimacy' when a fat body is touched and touches itself, displaces and disrupts the power of a purely visual register to control the arrangement, display and experience of female corporeality (see Rose 1993). By drawing upon Luce Irigaray's critical engagement with Merleau-Ponty's account of hands touching and the morpho-logic of the mucous, the chapter will challenge accounts of the fat female body that place her in a position of estrangement and distance from her varied materialities. Instead I will suggest that Jenny Saville's paintings of bodies are premised upon distinctly geographical relations of proximity and intimacy in ways which surprise and challenge our understandings of what a fleshy body can do.

It must be said that the use of visual (painted) representations of fat female bodies in this chapter should not simply be interpreted as privileging the visual over the tactile, as mentioned above. Instead, Jenny Saville's paintings are deployed firstly, because they draw attention to touching relations that cannot be seen by the viewer i.e. spaces between folds of fat, between legs and under breasts, and secondly, because the materialities of fat female bodies are presented in ways that challenge and evade a gaze that aims to control or eradicate fat from a body i.e. Saville's fat female bodies confront the viewer by holding and grabbing their fat and exceed containment by pushing out beyond the frame of the painting. In short, her paintings constantly remind us of the failure of the visual to capture fat in the same way that Merleau-Ponty's account of hands-touching ignores the sexed specificities of touching relations that are not always visible.

In the proceeding sections I develop the theoretical basis for intra-body touching by discussing the nature and significance of touch for Merleau-Ponty's account of perception in order to present Irigaray's critique and deployment of the mucous as an alternative to the phallocentrism of his account of hands touching. I will then provide more background concerning the broader ethos of Jenny Saville's work and thus present an account of intra-body touching with reference to two of her over-life-sized paintings, *Branded* and *Propped* (Figures 11.1 and 11.2). I then offer some conclusions regarding the potential of a feminist account of intra-body touching for making sense of (fat, female) bodies.

Merleau-Ponty and 'Hands Touching'

Merleau-Ponty's articulation of a 'flesh ontology' in his final and unfinished book *The Visible and the Invisible* (1968) builds upon his previous work, *The Phenomenology of Perception* (1962), which developed a phenomenological account of the 'lived body' or a 'body subject', understood as 'an always-already-incarnate subjectivity, as self inseparable from its embodiment' (Wylie 2007, 148). This account of the 'body subject' disputes a Cartesian notion of a singular body that 'experiences' or is 'experienced' in pre-given spaces. Instead the body is understood as the condition of our 'access' to space (Grosz 1994). He states that:

> The "here" of my body does not refer to a determinate position in relation to other positions or to external coordinates, but the laying down of the first coordinates, the anchoring of the active body in an object, the situation of the body in the face of its tasks. (Merleau-Ponty 1962, 100)

In his 1968 book, however, Merleau-Ponty shifts his focus from what Vasseleu (1998, 23) terms the 'physiological and psychological factors' in the process of perception to question the pre-discursive conditions through which perception emerges. In so doing, Merleau-Ponty (1968) introduces the idea of 'flesh' to describe a 'primary element' that does not belong to the mind or body, the subject or the object (world) and instead provides the 'conditions and grounds' (Grosz 1994, 95) for their distinction. Therefore, it is his way of articulating the immutability of the subject and the object in that 'flesh defines a position which is both subject (a subjective reality) and object (objectifiable for others), and also simultaneously a subject which is internally divergent with itself' (Vasseleu 1998, 26). Merleau-Ponty uses the example of the 'double sensation' of when one hand touches another in order to demonstrate the 'reversibility' of flesh 'if my hand, while it is felt from within, is also accessible from without, itself, tangible for my other hand' (Merleau-Ponty 1968, 133). In other words, the example of hands touching demonstrates the indeterminacy of what lies between seeing and being seen, and touching and being touched. Grosz (1994, 100–101) describes this relation as follows:

> In the double sensation my right hand is capable of touching my left hand as if the latter were an object. But in this case, unlike an object my left hand feels the right hand touching it. My left hand has the double sensation of being both the object and the subject of the touch. It is not the case that I have two contrary sensations at the same time (as one might feel two objects at the same time); rather each hand is in the ambiguous position of being capable of taking up the positions of either the toucher or the touched. If the double sensation makes it clear that at least in the case of tactile perception, the subject is implicated in its objects and its objects are at least constitutive of the subject, Merleau-Ponty wants to argue that such a model is just as relevant for vision.

This description of the double sensation draws attention to the ontological and empirical significance of body touching body relations. Merleau-Ponty also develops his account beyond the purely tactile register by interrogating the 'visual' and the relations between the visible (sensible) and the invisible (intelligible). His argument is not simply that to see is to be seen but to also proffer that what is seen, i.e. what is visible, also has the capacity to see. In short, 'the seer and the visible reciprocate one another and we no longer know which sees and which is seen' (Merleau-Ponty 1968, 151).

Merleau-Ponty's 'flesh' ontology, therefore, presents an account of perception that refigures the relationship between the subject and object. This is done by deploying what he names as the 'chiasm'. Chiasmic relations or intertwinnings between subject and object, vision and being seen by others, the visible and the invisible and touching and the tangible, are understood as being reversible in that they fall back on each other and yet at the same time are not the same thing. The relationship between them is not symmetrical. For example, Merleau-Ponty (1968, 147–148) states in relation to hands touching that 'reversibility is always immanent and never realized. My left hand is always on the verge of touching my right hand touching things, but I never reach coincidence; the coincidence eclipses at the moment of realization' (1968, 147–148). The chiasm and the reversibility of 'flesh', therefore, provide a useful way to consider the empirical and ontological significance of intra-body touching for fat female bodies. Whilst Irigaray and others, such as Young (1990) and Grosz (1994), are critical of Merleau-Ponty's deployment of visible and intentional touching relations and for ignoring the internal viscerality of bodies and the sexed specificities of bodies, he does offer an understanding of touch that is not necessarily hierarchical, possessive or controlling. Similarly, Jenny Saville's fat female bodies and materialities represent relations with and within fat that move beyond normative understanding of fat on a female body to be removed, concealed or sculpted (see Bordo 1993). Instead, in her paintings fat touches and is touched in ways which are indicative of fat subjectivities produced through relations of proximity and intimacy with fat rather than through distance and estrangement.

Luce Irigaray, Touch and 'the Mucous'

As already indicated, Luce Irigaray, in her book *An Ethics of Sexual Difference* (2004), critically engages with Merleau-Ponty's account of perception in order to develop an account of touch and touching that takes account of the sexed (female) specificities of the body. Broadly speaking, her work is premised on asserting and unpacking the phallogocentrism of Western philosophical traditions and their 'containment' of sexual difference and 'maternity' in so far as the potential of these for women as subject has not yet been fully realised (see also Irigaray 1985a, b).

The nature of her critical engagement with Merleau-Ponty (1968) is to reveal the implicitly feminine morphology or 'disembodied underside' (Grosz 1994, 103) of his account that helps to sustain his argument but that ignores sexual difference. Irigaray does this by focusing on his example of hands touching. She views the deployment of this example as reinforcing particular relations of dominance and hierarchy. For, she argues, it is presumed that one hand is given access to another and vice versa and that they are only irreducible to each other. She also takes issue with his easy comparison of the touching of hands to the two lips of the mouth that he uses when alluding to how the tangible and the visible are transposable to each other. Irigaray finds similarities with the problems of hierarchy associated with the hands touching example. She details how, in his description, one lip 'remains in or of its own sensible, another from which it will emerge, which it will see and to which it will stay tied as seer' (Irigaray 2004, 139). Irigaray here takes issue, firstly, with the implication that one lip will always 'see' or emerge from the other and, secondly, that all lips belong to the same sensible or body. Merleau-Ponty's description of lips and their relationship with the visible clearly ignores the sexed specificities of lips on a body. Instead, Irigaray (2004, 135) suggests we consider the following example of hands touching:

> The hands joined, palms together, fingers stretched, constitute a very particular touching. A gesture often reserved for women…and which evokes, doubles, *the touching of the lips* silently applied upon one another. A touching more intimate than that of one hand taking hold of the other. A phenomenology of the passage between interior to exterior. A phenomenon that remains in the interior, does not appear in the light of day, speaks of itself only in gestures, remains always on the edge of speech, gathering the edges without sealing them.

The specificities of lips (both the mouth and the labia) for the female body evokes a different articulation of the tactile than that proffered by Merleau-Ponty. For Irigaray, it is not possible to transpose the capacities of the tactile onto that of the visual because they function according to different logics; such differences relate to the sexual specificities of bodies. Moreover, her critique also centres upon the dominance of the visual over the tangible in Merleau-Ponty's account of perception. She questions the certainty with which the visible and the tangible can be considered for their reversibility and in fact suggests that the tangible is actually subsumed within the visible which, in Shildrick's (2001, 396) words, 'consigns women to passivity' and 'neglects the chiasm of bodily surface and depth'.

Instead of an active subject reaching out, as suggested by Merleau-Ponty's example of the two hands, Irigaray develops the idea of the 'mucous' to draw out the specificities of a sexually differentiated account of touch described as 'that most intimate interior of my flesh … a threshold of the passage from inside to outside; between outside and inside' (Irigaray 2004, 142). The mucous, therefore, exemplified by the lips of the mouth and labia, allows for other forms of touching relations whereby lips can 'touch themselves in her, within and inside women, without having recourse to seeing' (Irigaray 2004, 139). Both lips exemplify a morphology that expresses a 'lack of oneness' and does not have 'a graspable unitary form' (Robinson 2006, 101), thereby challenging Merleau-Ponty's (1968) description of the two lips belonging to the same sensible. She also refers to the ways that Merleau-Ponty fails to acknowledge the particular 'visible' and 'invisible' of 'intrauterine life' in his account of perception, described by Irigaray

(2004, 128) as 'the invisible of its prenatal life' or the 'insurmountable other of the visible not reducible to its invisible other side'. In summary, whilst Merleau-Ponty describes a body that can see itself touching itself and touch itself seeing itself, in contrast, Irigaray argues that the 'two lips' express 'a tangible intimacy which is experienced without reference to the visible' (Vasseleu 1998, 67).

In terms of framing an account of Jenny Saville's paintings, Irigaray's work on touch is significant for two main reasons. Firstly, the exposition of the phallogocentrism of particular scholarly accounts of the subject has resonance with Jenny Saville's approach to painting female bodies. Irigaray's approach involves a critical interrogation of Merleau-Ponty's account of the reversibility of touch for making sense of self-world relations. For Jenny Saville, this involves producing representations of the female nude that deliberately usurp the masculinist practices of painting women's bodies (see Nead 1992) and so our insecurities concerning the fat female form in contemporary Western society (see LeBesco 2004). Secondly, Irigaray's account of touch centres upon the sexed specificities of intra-body tactile relations by making explicit reference to the materiality of the female sexed body. This has involved developing a specific vocabulary for making sense of this specificity through the use of the mucous. Irigaray is not merely deploying the mucous as a metaphor for making sense of female embodied experience (see Robinson 2006). Instead, the mucous offers a different vocabulary and 'space-time' for the 'woman-as-subject' (Robinson 2006, 159) precisely because, in Irigaray's (2004, 93) words, it sets up 'the intimacy of bodily perception'.

In the analysis of Jenny Saville's paintings that follows, I demonstrate the ways that particular aspects of Irigaray's morpho-logic of the mucous can be used to frame the fat female body as something more than a temporary layer of flesh that is disavowed and in need of removal (Bordo 1993). This involves highlighting the materialities of intra-body touch, as exemplified with Irigaray's account of how lips touch, by drawing attention to the invisibilities of fat touching, for example the spaces that are created on bodies when fat folds and presses upon itself; acknowledging the existence of non-hierarchical relations of intimacy with fat that assert its existence rather its denial through the touching relation of grabbing; and by responding to the containment of fat and representations of female bodies more generally by fat-hating and phallocentric knowledeges and practices by demonstrating that fat and female bodies have capacities to break out and exceed those meanings placed upon them.

Jenny Saville and 'Over-life Sized Paintings'

Jenny Saville was born in 1970 and graduated from Glasgow School of Art in the summer of 1992. Following her graduation the British art collector Charles Saatchi offered her an exhibition at his London gallery and commissioned fifteen new works. This strand of her art practice culminated in 1997 in an exhibition of her 'over-life sized paintings' in the now celebrated 'Sensation' show of young

British artists. Her work has been likened to that of Francis Bacon, Lucien Freud and Willem de Kooning because of its explicit engagement with the human body and her techniques of working with paint to 'make flesh' (Rowley 1996).

Her 'over-life sized' paintings also draw comment for their place within an explicitly feminist politics concerning dominant representations of the female form (Robinson 2006). Working largely from photographs of bodies (her own and those in medical textbooks) she produced a series of large paintings of a female fleshy body that exemplified its surfaces, textures and orifices (Figures 11.1 and 11.2). In part, her work could be interpreted as an attempt to 're-present' the female body. For example, in an interview below she contextualises her work by explaining her frustrations of growing up in the AIDS obsessed and body conscious 1980s and finding no female artists in the art books she was looking at.

> Of course you start to ask "why not?" And: "Could I make a painting of a nude in my own voice?" It's such a male laden art, so historically weighted. The way women were depicted didn't feel like mine, too cute. I wasn't interested in admired or idealised beauty. (Saville interviewed in Mackenzie 2005, 6)

Art commentators have viewed her work in varied terms as 'depicting a clear-eyed and unromantic view of the average female form' (Roberts 2003, no pagination), as 'every woman's nightmare; vast mountains of obesity, flesh run riot, enormous repellent creatures who even make Rubens' chubby femme fatales look positively gaunt' (Henry quoted in Meagher 2003, 27) and as serving the interests of fat politics and fat pride because of her depiction of 'otherwise maligned and abjected bodies' (Milner 1997, 40). This uncertainty in the meaning and interpretation of Saville's art is interesting given the emphasis in this chapter on exploring touch as a way of exemplifying the physical and discursive ambiguity of the bodies in her work and of the fat female body more specifically.

What draws her work together, therefore, is an explicit interest in the materiality of bodily flesh. In an interview she says of flesh that it is everything: 'It's all things. Ugly, beautiful, repulsive, compelling, anxious, neurotic, dead, alive. And it's nothing. Eventually we expel ourselves. We rust away' (quoted in Mackenzie 2005, 6). The account of two of her paintings, *Branded* and *Propped*, that follows, begins from this description of bodily flesh as all encompassing. These two paintings have been selected because they present specific relations of touch that draw our attention to the fleshy materiality of the fat body but in a way that, as Saville herself mentions, does not stabilise the meaning of the fat body but opens up new possibilities for what a fat body can be. Moreover, the relations of touch presented in *Branded* and *Propped,* are comparable to the non-hierarchical and invisible touching relations that are central to Irigaray's deployment of the morphologic of the mucous.

Branded

The first painting is called *Branded* (Figure 11.1). The painting is of a woman who resembles Jenny herself. The woman looks down at the spectator with a gaze that emerges from half closed eyes. She seems disinterested. Her body fills the painting and is cut off at the top of the legs, which extend beyond the frame. A small sprig of pubic hair pokes out from the bottom of the frame. As she strains her neck back towards a wall she pushes the rest of her body out towards the viewer. Robinson (2006, 122) describes the bodies in Saville's over-life sized paintings as follows:

> we are invited into an intimate relation with their flesh, both by the scale of the figures [...] and by the way the scale of the figures butts up against the size of the canvas; they fill and exceed the field of vision in the painting, thus provoking a memory of a necessary closeness in order to make sense of the image.

Her breasts push out towards us, whilst simultaneously hanging down and casting a shadow on her stomach. Her left hand grabs a section of her stomach whilst her right hand rests on the top of her right thigh, gently resting upon folds of flesh that appear at her waist. She is thus touching herself. It is as if she is offering up her flesh to the spectator.

When situating *Branded* within the wider context of representing female bodies, I have found it useful to consider Goffman's (1976) work on the representation of gendered bodies in advertisements in the United States in the 1970s. In this work he classifies forms of touching relationships that are specific to men and women by collating hundreds of magazine adverts, determining patterns according to the gender of the body on display. He identifies what he terms the 'feminine touch', where women more than men are seen to caress or cradle the products they are advertising and the practice of women 'self touching,' which is 'readable as conveying a sense of one's body being a delicate and precious thing' (Goffman 1976, 31). He differentiates these forms of touching from what he calls a 'utilitarian' touch that 'grasps, manipulates or holds' (Goffman 1976, 31), which more aptly describes the grab or pinch of flesh in *Branded*. Whilst I am conscious of the problem with Goffman's (1976) dualistic reinforcement of feminine touch as 'passive' and masculine touch as 'active', I do feel that his work is useful for drawing attention to the various ways that bodies touch themselves. For example, in *Branded*, the woman's touch is active and forceful and has the effect of presenting the flesh to the audience almost as a challenge; this is mine, come and look. Moreover, the grab of flesh is reminiscent of contemporary advertisements for liposuction and other weight loss products, which often use (headless) images of bodies pinching flesh, usually in the area of the stomach or hip, to identify fat that can be removed or 'lost' by a procedure or a product. However, this time the effect of the pinch or grab is different. Saville makes the hand touch the body in a way that both challenges conventional representational practices of women touching themselves but also signals the ways that intra-body

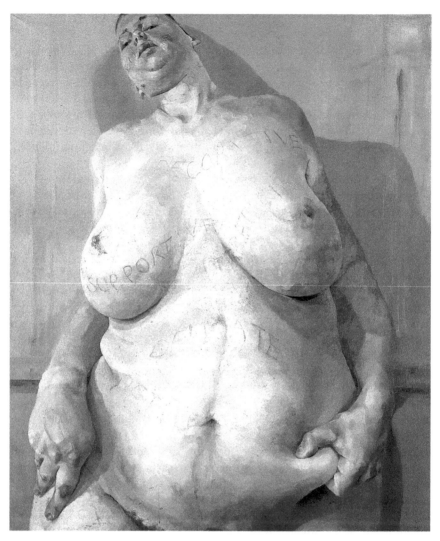

Figure 11.1 *Branded*, **by Jenny Saville, 1992 Oil on canvas, measuring 213cm by 183cm**

Source: courtesy of the Gagosian Gallery, image (c) Jenny Saville

touch is premised upon an intimate knowledge of the body and its capacity to 'act' in particular ways.

In previous work on fatness I have referred to these bodily capacities as indicative of the 'intra-action' of matter (Colls 2007). In short, this conceptualisation of bodily matter 'acknowledges the existence of specific body topographies, its textures and surfaces, and its capacities to create new spaces on the body' (Colls 2007, 363). The presentation of intra-body touch in *Branded*, therefore, is indicative of such materialities. It also challenges those geographical accounts of touch and the touching/touched subject, as discussed earlier, which imply that in order to touch, a body reaches 'out' to something other that its self or is touched by that 'other'. Instead, touching or grabbing our own body exemplifies how bodies can be always and already touching. In some ways this is reflective of the promise of Irigaray's (2004) notion of the mucous deployed in this instance as a way of acknowledging the ambiguity of touch. The mucous, in short, 'escapes control, not being subject to the kind of voluntary slippage by which the touching hand becomes the touched' (Grosz 1993, 51). Whilst the hand grabs the fat on the body in a way that could indicate a relationship of domination or violence, it does not take control of the flesh in order to police its materiality; instead, the fat affirms itself in a way that emphasises its place upon the body.

Written on Saville's body in *Branded* are the words *Supportive, Irrational, Delicate, Decorative, Petite*; all of which encompass the qualities of traditional feminine beauty and psychology (Figure 11.1). These words are etched upon the skin, touching the body in the same ways that discourse and meaning are placed upon women's bodies in an attempt to define them. However, the discursive power of these words to fix their object of definition is disrupted by the materiality of the fat body beneath, which seems to contradict both the meaning of the words and the conditions of their deployment. For example, the words 'petite' and 'delicate' clearly fail to adequately describe the size of the body represented, but do describe the desirable size and comportment of a female body within a weight-obsessed Western society. In one account of Saville's work, Meagher (2003, 24) comments on the way that her paintings provide 'a provocative site for the emergence of an aesthetics of disgust that can propose new modes of thinking about feminine embodiment'. For her, the paintings by Saville encompass an overtly feminist project that both challenges the conventions of representing a female nude and also forces spectators to confront their feelings and sources of disgust in relation to the body in the paintings. She states:

> By materialising the abject body, Saville reveals what lurks in the feminine imagination. That is to say, by representing a specific idea of femininity, she speaks to the disparity between the way that many women feel about their bodies and the reality of how those bodies are perceived by others. (Meagher 2003, 34)

However, by focusing on the function of disgust in relation to the consumption of the paintings Meagher (2003) risks, firstly, making sense of these bodies through an interrogation of a negative relationship with fat as 'disgusting' and, secondly, ignoring the significance of the represented materialities of fat bodies, for example how bodies touch and are touching. Significant to this painting are the presence of folds of bodily flesh that, like the lips of the labia for Irigaray (2004), demonstrate the irreducibility of subject and object. Indeed, Paterson (2007, 106), when commenting upon the forms of the body in digital art remarks that, 'folding is immanent, brings the outside in, collapses distances, and makes presences felt'. The folding of fat flesh in this way creates new tactile spaces upon the body, which are constantly 'coalescing' (Bordo 1993) and changing according to the comportment of the body itself. They form spaces that cannot necessarily be seen but can be felt as folds touching folds or as hands, fingers, or other parts of the body, touching, pressing and mingling between them. To return to Irigaray's (2004) critique of Merleau-Ponty's (1968) description of the lips and the visibility of touch we can liken the spacing of folds of fat on the body, displayed in Saville's painting *Branded*, to the potentialities of the 'tangible invisible' of intrauterine life and the lips of the labia. Vasseleu (1998, 67) describes the tangible invisible as 'the body as a positive reserve, a vitally constituted dimension, an adherence to indetermination rather than the surfacing of an unpreventable interior'. Fat materialised on the body as folds in this way, therefore, is not simply the abject 'other,' nor is it a hidden space inside the fold that cannot be seen. Instead, like the mucous, the intra-body touching of fat folding encapsulates 'an interior which could not be more intimately me, yet which evades my mastery' (Vasseleu 1998, 68).

Propped

In the second painting entitled *Propped* (Figure 11.2), we see the same woman as featured in *Branded*. She is sitting on a stool, apparently naked except for a pair of white sling back shoes. She leans slightly forward and her feet are hooked around the pole of the stool. The woman looks down upon us with a somewhat defiant gaze; her head edging out of the frame of the painting but maintaining direct contact with the viewer of the painting through one eye. The woman grabs her thigh with both hands, reminiscent of the grabbing of flesh featured in the previous painting. The fingers reach into her thighs with force as we see her nails dig into the skin, her fingers make indentations in the surface of her legs. Her breasts hang down towards her lap and are pressed upon either side by each forearm. The subject's stomach is hidden by her arms and hands. The seat of the stool is also absent, buried under the thighs and yet supporting it. Her legs are pressed together making it impossible to see between them.

The pressing together of the woman's thighs, which hints at and yet hides that which is hidden between her legs, can be likened to Irigaray's (2004) assertion of particular touching relations that are 'not visible'; those that are denied in Merleau-Ponty's (1968) account of perception. *Propped* hints at the internal touching of the lips of the labia and indeed the space that is fleshy thighs rubbing and pressing together (Figure 11.2). Moreover, it is also interesting to note Saville's particular choice of comportment of the woman's breasts in this painting. As they hang down they are also pressed into the arms of the woman such that all parts of the body are touching in a relation of breasts, arms, stomach, hands, thighs. This meeting of breast, fat and folded skin highlights what Young (1990) would call the 'fluidity' or 'instability' of women's bodies, which does more than simply assert its biological difference. Instead, she argues, the materiality of female bodies

Figure 11.2   ***Propped*, by Jenny Saville, 1992. Oil on canvas, measuring 213cm by 183cm**

Source: courtesy of the Gagosian Gallery, image (c) Jenny Saville

can disrupt the privileging of vision over touch in the objectification of women's bodies by men. To exemplify this Young (1990, 195) describes the materiality of breasts as follows:

> Without a bra, a woman's breasts are also deobjectified, desubstantialized. Without a bra, most women's breasts do not have the high, hard pointy look that

phallic culture posits as the norm. They droop and sag and gather their bulk at the bottom.

In the *Propped* painting Saville consciously re-presents the breasts of a fat female body in ways that differ from the traditional or phallocentric conventions of representing the female body. Indeed, the 'more than one' of intra-body touch can be clearly demonstrated through the pressing together of flesh, which creates new spaces upon the body, both visible and invisible, to the spectator. Both of these relations of intra-body touch resonates with the mucous, as the folds of *Branded* do (Figure One), in providing a setting for the intimacy of female and fat bodily perception. Written over the surface plane of the painting, but not on the woman's body, is a quote by Irigaray taken from her book *Speculum of the Other Woman*. It reads:

> if we continue to speak in this sameness – speak as men have been doing for centuries we will fail each other. Again words will pass through our bodies above our heads – disappear, make us disappear. (1985a, 205)

The quote is indicative of Irigaray's (1985a) wider project to expose the phallogocentrism of philosophy, and in particular the language of philosophy, which contains and denies the female subject, as I have discussed above. It is written back to front over the surface of the picture rather than on the body as in *Branded*. (In a gallery setting a mirror is placed behind the viewer so that they can see the painting and text in the mirror.) The quote also demonstrates Jenny Saville's most explicit reference to the feminist ethos of Irigaray's work. As with Irigaray's (2004) identification of the denial of a female morphology in Western philosophy, Saville has struggled to find herself or her body in conventional paintings of the female nude. She, therefore, could be described as producing her own 'morpho-logic' by re-presenting the female fat body on her own terms. As she states, 'I want the bodies to be located in the painting […] I want it (a story) in the body. I don't want to illustrate, I want it to be what it is' (quoted in Schama 2005, 127).

Isabel Wallace (2004, 79) comments in depth about the significance of the painting as being part of a process with three stages, as, 'first, a sensuous construction and experience of the female body; second, surgical inscription of the letters; and third, the production of a specular double'. In her analysis of the painting Wallace (2004) concentrates on working through the ways that *Propped* is indicative of Irigaray's argument concerning the masculinist specular regime that governs women's bodies. In short, the written text on the painting contains the body and mediates our relationship with it. We turn our back on the woman as we look in the mirror, and we then see a re-presentation of the female nude in a reflection of Saville's painting. Saville's mirror both exposes us to, and implicates us in, the masculinist viewing position that has fixed and framed fat bodies and women's bodies more specifically. As Irigaray describes (2004, 143), the mirror 'gives access to another order of the visible. Cold, icy, frozen-freezing […] I see myself in the mirror as if I was other'. However, Saville's *Propped* body does not disappear, as Irigaray (2004) warns in relation to language. Whilst Irigaray's words do contain the body beneath them, 'propped' by language as the title suggests, the body itself defies this containment by challenging those phallogocentric conventions that have been implicit in the body's containment. Irigaray's words function as a warning of what has been, while the fat body beneath, exemplified by the intra-body touching relations of breasts, fat and folds of skin, suggests what is possible outside of the processes of containment.

Conclusion

> The organ which has nothing to show for itself also lacks a form of its own. And if woman takes pleasure (*jouit*) from this incompleteness of form which allows her organ to touch itself over and over again, indefinitely, by itself, that pleasure (*jouissance*) is denied by a civilisation that privileges phallomorphism. (Irigaray 1985b, 26)

I begin this conclusion with a quote from Irigaray (1985b) that forms part of a longer description of the female genitals. The quote illustrates the way that, for Irigaray, a phallogocentric logic polices the form of the female body, i.e. that it has nothing to show for itself because it is indivisible and invisible, unlike the phallus. It also suggests the potential offered by the 'incompleteness' of the female form, or mucous, for female pleasure and for developing a 'morpho-logic' with which to make sense of woman as subject. Within this chapter I have argued that this dual orientated argument can be used when considering the presentation of a fat female morphology in the over-life sized paintings of Jenny Saville. In particular, by drawing attention to the 'intra-body touching relations' of fat upon, within and inside female bodies, new and previously unknown spatialities of fat have been discerned. Indeed, the specific morpho-logic of Irigaray's notion of the mucous has helped to highlight the corporeal intimacy and proximity of touch, which are exemplified by the grabbing and folding in *Branded* and the breasted corporealities and the re-working of the visual distancing of fat in *Propped*. In Shildrick's (2001, 388) terms, this ambiguity is a manifestation of 'a difference that forces us to question the ontological and ethical status of distinction'; a distinction that has previously placed fat as an unwanted 'other' upon bodies, and which does not allow for a woman to have a relationship with fat beyond that of estrangement and disavowal.

In this chapter I have also endeavoured to highlight the role and value of feminist critical and theoretical work on the body for making sense of the (fat) female subject. Specifically, Irigaray's critical account of Merleau-Ponty provides a critical lens to 'gaze' upon his incredibly useful body of work on the formation and constitution of the subject and subject-object relations. A way of illustrating this potential can be seen in the usefulness of Irigaray's notion of 'mucous'. The mucous should not be understood just as a linguistic tool or a metaphor for uncovering an-other empirical hidden. Indeed, Robinson (2006) warns against this tactic in the context of using Irigaray's ideas to produce art, and suggests that developing a 'morpho-logic' focuses on the functioning of morphology rather than its meaning. When thinking about the spatialities of bodies and their materialities it is useful, therefore, to consider the nature of the 'morpho-logic' being created or sustained through our engagement with and application of particular philosophical bodies of work. It is important to consider the material, sexed, sized or otherwise, specificities of bodies that are significant in order to avoid reproducing exclusionary and phallogocentric accounts of who/what that subject might be, or is capable of.

Finally, I wish to comment on the particular relation of touch, or what I have called 'intra-body touching,' that has been used in this chapter in order to explicate the specificities of fat touching and being touched on, in and across bodies. I feel that acknowledging these relations of touch can be productive for thinking empirically about the spatial form of the fat body. As I intimated at the beginning of the chapter, work on touch in academic geography has not yet fully engaged with the ways that bodies are always and already touching themselves according to the specificities of their materiality. An attentiveness to intra-body touching would mean reconsidering the relationships between subject-object, self-other and interior-exterior relations as they are present within and upon specific bodies. For example, in this chapter, intra-body touching has been useful for 'uncovering' the specific morpho-logic of a fat, female body. Moreover, Shildrick (2001) demands that we reconsider the place of touch in the production and reception of all non-normative subjects. The everyday malignment of fat bodies (and fat *on* bodies) has had the result of removing fat from the body both literally and discursively, through its denigration in language and physical intervention. Instead, the presentation of fat bodies in this chapter, particularly through the visual strategies deployed by Jenny Saville, highlights the productive potential of relations of intimacy and proximity to, and as, fat bodies through touch. The ethical and political potential of this form of questioning of fat bodies is important for a reformulation of the 'fat' subject, and other non-normative subjects, which are produced more often that not in a relation of estrangement and distance from themselves and other bodies.

Acknowledgements

I would like to thank Mark Paterson and Martin Dodge for their feedback and suggestions. I would also like to thank James McKee at the Gagosian Gallery, New York for her help in obtaining permission to use images of Jenny Saville's paintings in this chapter. Finally, gratitude as always, to Ben Anderson for his 'hope' and support.

References

Bordo, S. 1993. *Unbearable Weight: Feminism, Western Culture and the Body.* Berkeley, CA: University of California Press.

Colls, R. 2007. Materialising bodily matter: Intra-action and the embodiment of 'fat'. *Geoforum*, 38, 353–365.

Davidson, J., Bondi, L. and Smith, M. 2005. *Emotional Geographies.* Aldershot, England: Ashgate.

Dixon, D. and Straughan, E. 2010. Geographies of touch/touched by geography. *Geography Compass*, 4(5), 449–59.

Goffman, E. 1976. *Gender Advertisements.* London: Macmillan.

Grosz, E. 1993. Merleau-Ponty and Irigaray in the flesh. *Thesis Eleven: Critical Theory and Historical Sociology*, 36, 37–59.

Grosz, E. 1994. *Volatile Bodies: Toward a Corporeal Feminism.* Bloomington, IN: Indiana University Press.

Hayes-Conroy, A. and Martin, D. 2010. Mobilising bodies: visceral identification in the Slow Food movement. *Transactions of the Institute of British Geographers*, 35, 269–81.

Ingold, T. 2004. Culture on the ground: the world perceived through the feet. *Journal of Material Culture*, 9(3), 315–40.

Irigaray, L. 1985a. *The Speculum of the Other Woman.* Translated by G.C. Gill. Ithaca, NY: Cornell University Press.

Irigaray, L. 1985b. *This Sex Which Is Not One.* Translated by C. Porter and C. Burke. Ithaca, NY: Cornell University Press.

Irigaray, L. 1991. *The poverty of psychoanalysis, in Luce Irigaray: Philosophy in the Feminine*, edited by M. Whitford. London: Routledge, 79–104.

Irigaray, L. 2004. *An Ethics of Sexual Difference.* Translated by C. Burke and G.C. Gill. London: Continuum.

Lea, J. 2009. Becoming skilled: The cultural and corporeal geographies of teaching and learning Thai Yoga massage. *Geoforum*, 40(3), 465–74.

LeBesco, K. 2004. *Revolting Bodies? The Struggle to Redefine Fat Identity.* Boston: University of Massachusetts Press.

Longhurst, R., Johnston, L. and Ho, E. 2009. A visceral approach: cooking 'at home' with migrant women in Hamilton, New Zealand. *Transactions of the Institute of British Geographers*, 34, 333–45.

Mackenzie, S. 2005. *Under the skin.* The Guardian, 22nd October.

Macpherson, H. 2009. Articulating blind touch: thinking through the feet. *Senses and Society*, 4(2), 179–92.

Meagher, M. 2003. Jenny Saville and a feminist ethics of disgust. *Hypatia*, 18(4), 23–41.

Merleau-Ponty, M. 1962. *The Phenomenology of Perception.* Translated by C. Smith. London: Routledge and Kegan Paul.

Merleau-Ponty, M. 1968. *The Visible and the Invisible.* Translated by A. Lingis. Evanston, IL: Northwestern University Press.

Milner, C. 1997. *The arts: Bring on the blubbernauts*. The Sunday Telegraph. 14th September.

Nead, L. 1992. *The Female Nude: Art, Obscenity and Sexuality.* London: Routledge.

Obrador-Pons, P. 2007. A haptic geography of the beach: naked bodies, vision and touch. *Social and Cultural Geography*, 8(1), 123–41.

Paterson, M. 2007. *The Senses of Touch: Haptics, Affects and Technologies.* Oxford: Berg.

Roberts, A. 2003. *The female gaze*. The Observer. 20th April.

Robinson, H. 2006. *Reading Art, Reading Irigaray: The Politics of Art by Women.* London: I.B. Taurus.

Rose, G. 1993. *Feminism and Geography: The Limits of Geographical Knowledge.* Cambridge: Polity.

Rowley, A. 1996. On viewing three paintings by Jenny Saville: rethinking a feminist practice of painting, in *Generations and Geographies in the Visual Arts: Feminist Readings*, edited by G. Pollock. London: Routledge, 88–109.

Schama, S. 2005. Interview with Jenny Saville, in *Jenny Saville Catalogue*, edited by J. Saville. New York: Rizzoli, 124–29.

Shildrick, M. 2001. Some speculations on matters of touch. *Journal of Medicine and Philosophy*, 26(4), 387–404.

Vasseleu, C. 1998. *Textures of Light: Visions and Touch in Irigaray, Levinas and Merleau-Ponty.* London: Routledge.

Wallace, I. 2004. The looking glass from the other side: Reflections on Jenny Saville's Propped. *Visual Culture in Britain*, 5(2), 77–91.

Whitford, M. 1991. *Luce Irigaray: Philosophy in the Feminine.* London: Routledge.

Wylie, J. 2007. *Landscape*. London: Routledge.

Young, I.M. 1990. *Throwing Like a Girl and Other Essays in Feminist Philosophy and Social Theory*. Bloomington, IN: Indiana University Press.

Chapter 12

Touched by Spirit: Sensing the Material Impacts of Intangible Encounters

Sara MacKian

Introduction

> Reality is merely an illusion, albeit a very persistent one. (Albert Einstein,
> quoted in Tyler 2009, no pagination)

Since time immemorial the human relationship with spirit, whether all-embracing
or outright hostile, has been a key part of how people sense and make sense
of what they perceive as their own personal realities. The archaeology of our
social and physical worlds gives testament to the tangible imprint of spirit and
spirituality through time. Spirit touches our world and leaves its mark, whether
we believe in it or not.

The dominant cultural expression of human relations with spirituality is in the
form of religion (Hay and Nye 2006). From the stone boulders of Castlerigg to the
cathedral of Notre Dame, the Western world bears the physical imprint of religious
beliefs and practices. Whole societies are defined and recognised by their religious
identities – 'Christian enclaves', 'the Islamic world', the 'Amish community' or
disputes over the route of Protestant Orange Parades through largely Catholic areas
in Northern Ireland. In these we witness very tangible articulations of the human
relationship with the spiritual, in the form of buildings, church-going, ritualised
prayer practices or particular styles of dress. Yet lost and hidden beneath an
over-familiarity with such explicitly religious imprints in our social and physical
landscapes is the invisible 'spirit' which still possibly lies somewhere at its heart.
In part this reflects the demise of the supernatural worldview. Many believe the
disenchanted modern Western world no longer has need of spirits, angels and divine
beings. Since the age of Enlightenment, numerous social thinkers have suggested
that religious practices, rituals and superstitions would decline in importance with
the path of modernity (Norris and Inglehart 2004). Indeed, the tendency towards
secularisation in the late twentieth-century appears to confirm such predictions.
However, with the apparent demise of organised religion, we have simultaneously
witnessed the proliferation of 'personally adopted' and 'eclectic' spiritualities
(Hay and Nye 2006), with spiritual exploration being divorced from traditional
religious contexts and opening up uniquely personal opportunities to touch the
divine. This reflects the broader tendency towards reflexive individualisation,

social atomisation and detraditionalisation in late modernity (Giddens 1994, Beck 1994, Beck and Beck-Gernsheim 2002). A core feature of this latest spiritual evolution has been a return to supernatural beliefs and otherworldly interests as a means for 'spiritual seekers' to forge a meaningful relationship with 'spirit' (Partridge 2004, MacKian 2012).

The collapse in Christian church attendance in Britain at the close of the twentieth century coincided with a sixty per cent rise in reported spiritual experiences; patterns reflected across the European continent and Australia (Hay and Nye 2006). We are told that some twenty-seven per cent of the British public believe in reincarnation, twenty-two per cent believe in astrology and even fifteen per cent believe in the power of fortune telling (Spencer and Alexander 2009); forty seven per cent of Canadians believe in ghosts (Ipsos Reid 2006); Italians, Germans, Swedes and Icelanders have all reported experiences of contact with spirit (Hart Wright 2002); and three out of every four people in the USA believe in paranormal phenomena, including extrasensory perception, spirits or spiritual healing (Moore 2005). As a result, a generation of 'spiritual but not religious' individuals has emerged (Roof 1999), its exponents characteristically distancing themselves from institutional versions of religion, preferring to develop their own 'reflexive spirituality' in an attempt to find meaning and purpose in their lives. A growing part of this more reflexive approach to the spiritual involves an active engagement with 'sprit' as an otherworldly agent.

This contemporary alternative spiritual landscape offers a unique opportunity therefore to explore the less visible manifestations of earthly relations with spirit. Not everyone needs to sit in a church pew to get in touch with the divine, as there are other less physically obvious ways in which individuals are touched by spirit. In this chapter I suggest that the role of spirit should be revisited to reflect this growing popularity amongst alternative spiritual seekers to directly touch and be touched by the putatively intangible divine. In particular, by exploring this relationship through the lens of touch, we can gain deeper sociological insight into the way in which the touch of spirit is felt and seen in the modern world. Despite this apparent spiritual turn, critical social science has so far failed to grasp the significance of the re-enchantment of people's relationship with spirit, with leading commentators continuing to insist that 'God is dead' and there is no place for the supernatural in today's modern world (Bruce 2002). I suggest instead that there is actually an intimate – and very touching – relationship between the two.

I begin with some background to the empirical study underpinning the chapter, and then go on to outline – and in turn to challenge – the prevailing sociological thesis, grounded in the Weberian approach, that the modern world is a 'disenchanted' one with no place for spirit. Drawing on fieldwork evidence of the impact of spiritual encounters in participants' everyday worlds, I move on to suggest these give us 'new imaginations' of the world and how that world works (Dewsbury and Cloke 2009). In these imaginations emphasis is placed on the way in which points of contact – or moments of touch – between spiritual practitioners and spiritual energies map across the contours of everyday life, marking the interface between a world we think we know through the senses and an 'otherworld' less readily available. Contact has its roots in 'Tactus' the Latin word for 'touch' or 'sense of feeling', and is therefore particularly apposite here; for it embodies both the act of physical touching and the 'sense of *feeling*' which are key to the notion of spiritual touch I am exploring. Furthermore, given its association with distant communication – to be 'in contact with' – grounding my understanding of touch in the language of contact hints also at the extrasensory nature of spirit's 'touch'. Through attending to the 'touch of spirit' in moments of contact between 'this world' and 'the other', we can uncover new realities suggesting the imprint of spirit in our material world may be more persistent than the disenchantment thesis maintains.

Background to the Study: 'Reaching for the Unseen'

The empirical research underpinning this chapter began as an exploration into contemporary alternative spirituality (MacKian 2012). Early on in my investigations I discovered that it was common for these spiritual seekers to speak of being 'touched by spirit', and for feelings of security, love, benevolence and 'vastness' to accompany such sensations. This relationship with spirit was not an optional part of what spirituality meant for them, but a central feature. For many, it was only when they met spirit in the form of guides or angels that their spirituality became 'real' as these encounters served as evidence of a connection between, and transgression across, the boundary between this world and the world of spirit (MacKian 2010).

My interest grew in the worlds inhabited by those who routinely sought such a relationship with the otherworldly, and I explored a group of people who exhibited diverse interests and practices, but who nonetheless reflected the same broad approach to and understanding of their spirituality. For them spirituality was not housed within any particular religious tradition and did not require any regular, dedicated ritual or practice. It did, on the other hand, involve contemplation of what lies beyond our apprehension of the physical and material world, and much of this contemplation led to a sense of being connected spiritually (and through spirit) to something 'more than' that physical world. Therefore my research participants included, amongst others, spiritual mediums, angel healers and Reiki masters (see MacKian 2012 for a full account of the research process and methodology). 'Spirit' to them might be defined differently, as the spirit in nature, the spirit of angels, spirits of the deceased, or the universal spirit; but at the heart was the consistent desire to touch and be touched by something.

In most academic accounts of contemporary alternative spirituality there is a distinct lack of attention given to the 'mystery' or 'otherworldliness' which seemed to me to be an essential part of such spiritualities. There is even less account given of how this otherworldliness may affect the everyday world as we know it. Indeed, you will be hard pushed to find any serious or explicit engagement with mystery, enchantment and its relation to the quotidian and mundane. Numerous authors stress the material and earthbound orientation of contemporary spirituality, focusing on consumer behaviour in the spiritual marketplace, or giving accounts of isolated individualised and narcissistic self-worshippers (Bruce 2002, Carrette and King 2005). The emphasis is on self-contemplation, the inner life and the divine within (Heelas 2008); and Voas and Bruce (2007, 51) go so far as to claim this spirituality has 'little to do with the supernatural or even the sacred; it appears to be a code word for good feelings'.

Any mention of 'spirit' as part of these spiritualities is apparently overlooked, and you will not even find the word 'spirit' listed in the indexes of most of the leading texts in the field. If there is an attempt to engage with the more mystical aspects of spiritual pursuit the usual approach is to suggest they pertain to specific times and places, such as yoga retreats (Hoyez 2007), the privacy of very personalised rituals in the home (Holloway 2003), or the intimacy of the treatment room (Heelas and Woodhead 2005). However, it should not be forgotten that spirituality affects *all* aspects of life, and can have profoundly world-changing effects as a result. My fieldwork suggested that practitioners routinely incorporated otherworldly connections in their daily lives, and this everyday imprint of spirit's touch in the modern world is being overlooked in mainstream sociological discourse.

The Importance of Touch

Is the omission of spirit from the discourse because it is hard to get a handle on precisely what spirit might be and how one might measure its impact in everyday life? 'Spirit' is commonly assumed to be related to that which is *not physical*, those characteristics which are *not of the body*, but relate to deeply held beliefs or feelings, or even another supernatural world entirely. Spirituality accordingly is a concern with these non-physical elements, with these deep feelings and beliefs, and maybe even the ethereal or supernatural, rather than the more physical aspects of life, thereby involving potentially greater challenge to social scientific approaches.

Perhaps a more useful approach is to focus not on the ontological basis of what we are dealing with, but rather on the implications of being in the world in relation to certain core beliefs about spirit. Social science seems more willing to grasp the 'absent divine' in more earthly, concrete contexts. Hetherington (2003, 1940) uses the example of holy relics as *praesentia* to suggest the 'experience of mingling' between human and non-human – presence and absence – can be apprehended to witness the 'involvement of the absent Other within the material presence of social life'. With the material presence of holy relics, the absent divine is touched both directly – with fingers tracing the contours of the material artefact, and indirectly as a form of 'tacit understanding' about what those relics represent. However, as stated previously, the act of touching the 'supernatural' in contemporary spirituality – either directly or indirectly – has been neglected to date in scholarly analysis.

A range of 'present absences' (Wylie 2009) – from haunting memories in particular places experienced by almost everyone, to the more otherworldly hauntings of 'ghosts' and other inexplicable happenings – exist all around us. Space is 'shot through' with 'praesentia' – intimate and touching encounters with the 'presence of an absence' (Hetherington 2003, 1937), and holy relics offer a tangible way of encountering that. However, those encounters which lie at the more otherworldly, and less culturally acceptable end, continue to challenge the rationalist scientific mind. Even literature purportedly dealing with the place of the spiritual in the modern world fails to take account of the praesentia of spirit, dismissing any direct otherworldly interactions as 'paranormal' rather than 'spiritual' (see, for example, Heelas 2008).

In this chapter I suggest there is an alternative reading if we focus on moments of contact with spirit in the everyday lives of practitioners. As I will show, touch – as an act of connection and contact – is a fundamental feature of these encounters:

> I felt these wings wrap around me, just like an angel, and I was literally lifted
> from the bed and for the first time in days I felt at peace. I just knew I was being
> looked after by spirit. (Kathryn, research participant)

Sometimes practitioners 'reach out to' spirit, at other times – like for Kathryn – spirit 'comes knocking'. The *implications* of such moments of 'being touched' would appear to be that the spiritual itself is given agency. It is the centrality of touch to this agency that I consider in this chapter through the worlds and experiences my participants shared with me.

Touch is of course far from straightforward. Whilst we might most commonly associate touch with something that happens primarily physically, and is the result of things and us being in 'a particular position, at a particular temperature' and possessing 'certain textural qualities' which are sensed through the act of touch (Paterson 2009a, 130); as Paterson goes on to state, touch can also be something which is mediated and illusory, extended in less direct ways by 'haptic technologies', changing the ways in which the body 'talks' and is 'addressed' (Thrift 2004). Furthermore, to talk of 'being touched' can extend the vagaries of touch even wider into the emotional, the intangible and the metaphoric.

Notwithstanding the complexities of touch, the centrality of the tactile to what it means to be and feel human appears largely undisputed. To touch and be touched, and to sense and respond to touch not just physically but also emotionally and even spiritually, is part of what it means to be human in relation to everything around us. The things we are touched by shape our worlds, marking out the contours of the tangible and intangible 'realities' we live in. As a result: '[l]ife without touch is almost unimaginable' (Paterson 2009a, 129). Furthermore, evidence suggests that touch has profoundly therapeutic benefits. Of particular interest here, this 'touch' need not consist of direct physical contact to achieve results. The potentially therapeutic effect of touching the human biofield – or 'aura' – has been the subject of scientific investigation, and 'significant effects' have been reported from the energy medicine 'therapeutic touch' (TT) (Monzillo and Gronowicz 2011). Also known as 'non-contact therapeutic touch', the treatment involves the practitioner placing their hands near to a patient – rather than touching the body – to manipulate the energy field and promote therapeutic benefits. TT has been demonstrated to improve recovery for postoperative patients (Coakley and Duffy 2010), reduce pain, depression and sleep disturbance in patients with chronic pain (Marta et al 2010), and enhance psychological processes to promote optimum health for psychiatric patients (Vickers 2008). Despite an overwhelming focus on the body in discourses of touch, the success of TT is indicative of the importance of touching other levels of existence and consciousness (like Kathryn's experience in her hospital bed), not just the body.

To be spiritually active has also been shown to have health promoting qualities (Miller and Thoresen 2003, Levin 2003), and reinforces a sense of ontological security (MacKian 2012). Contemporary alternative spiritualities in particular have been described as ultimately 'therapeutic' (Horowitz, quoted in Winston 2009; see also Heelas and Woodhead (2005) on the 'holistic milieu'):

> hundreds of thousands of people dabble in all kinds of esoteric healing therapies, varying from mesmerism to spiritual healing, from Reiki to neo-shamanism or reincarnation therapy. (Vellenga 2008, 331)

The popularity of these distinctly esoteric, and even otherworldly, forms of treatment is possible testament to the fact that such therapies often involve more time, care and attention being given to the individual than with standard biomedical treatment regimes. A typical esoteric consultation may involve taking an extensive personal history, a considerable amount of relaxing hands on 'treatment' of some sort, and quite often this is accompanied by an element of counselling where it might be deemed appropriate. These elements of personal time, attention and extended dialogue in themselves have beneficial therapeutic impacts, even if actual 'cures' are not part of the prescription.

However, such mediated healing opportunities were not the only type of personal healing reported to me. A number of research participants perceived a more direct healing intervention from spirit. Richard was one participant who described his experience of suffering a mental breakdown followed by physical illness and an extended period of absence from work. During this time he was visited by the 'healing angel' Archangel Raphael:

> For me it was like spirit pulling me up short and giving me a rap across the knuckles. And then a much needed tonic! (Richard)

The place and role of touch in relation to the spiritual is therefore particularly interesting, as it lies at the intersection of these overlapping therapeutic spheres. The moments of touch I discuss in this chapter would on the face of it appear to primarily fall into the category of 'non-contact' touch, like Kathryn's and Richard's angel experiences, rather than for example hands-on healing sessions. However, I also stress that by locating the *everyday spaces* in which these moments of touch occur, it is possible to identify associated physical consequences in the material world. These are understood by practitioners to be the direct result of these touching encounters. Therefore, as well as communication with spirit, there is I argue, also 'contact' made between spirit and the material world.

Of course this argument asks the reader to suspend judgement over the 'paranormal' claims being made by my participants, and this was my stance for the fieldwork. After all:

> experiences do not have to be 'explained', but simply 'understood' as the way
> of experiencing the world that is natural and unremarkable, strange only to the
> outsider. (Knibbe and Versteeg 2008, 49)

Foltz (2000), in participatory fieldwork with witches, recognised the transformative effect of spiritual practice on the women's everyday lifeworlds, and in part it seems through the process of conducting the fieldwork, she was touched herself and began to feel a deeper sense of connection and awareness of her own place in the world. Understanding living in the world *as experienced* by those who are being researched is therefore a very valuable thing for researchers, but something which is not easily achieved. Paterson (2009b, 779) has suggested that ethnographic fieldwork sensitive to the 'haptic geographies' of touch, emotion and unusual bodily sensations can open up new possibilities for seeing, hearing and representing 'the taken-for-granted, the left unsaid, and sometimes the ineffable or tacit knowledges that emerge through encounters'. I believe my approach to these encounters does precisely that, in opening up new possibilities for sensing and representing these touches from otherworlds. Before analysing them in detail however I wish to look a little closer at the thesis of disenchantment which appears to have prevented such stories from being heard in critical social science studies of contemporary spirituality; and to offer an alternative view that allows us to embrace the otherworldly as a legitimate focus of study.

Disenchantment and the Death of Spirit?

> The 'supernatural has departed from the modern world.' (Berger 1970, 1)

The prevailing view has long been that people live in a disenchanted world in the West, where there is little room or need for anything that cannot be verified by rational means. The argument is that modernisation brought with it rationalisation, bureaucratisation and a steady erosion of the conditions needed to experience the world as some kind of enchanted and enchanting 'creation' (Wirzba 2003). Since Max Weber, an enduring myth of the history of modernity has characterised it as an inevitable 'demise of the social significance of the supernatural worldview' (Partridge 2002, 237); hence the apparent dismissal of spirit from spirituality. In this 'disenchanted' world, stripped of magic and meaning (Schneider 1993), there is supposedly little room for mysticism and wonder, as all areas of human experience have purportedly been 'conquered by' science and rationality (Jenkins 2000). The culmination of this for Weber was to trap individuals within an 'iron cage' of rational control. This narrative of a disenchanted modernity, inspired by Weber's iron cage, leaves only a 'cold and uninspiring world' (Bennett 2001, 60), which prevents people from contemplating what might lie beyond the lacklustre façade. With this general disenchantment, knowledge and perception of what might be widely accepted as 'real' is generally restricted to what people can experience with their five senses, and what can be measured empirically and 'known' rationally. In relation to the spiritual or religious, the argument is that disenchantment led to secularisation and the decline of mysticism, and removed the possibility for the individual to directly experience, know or be touched by the divine. The 'removal of God' from this world, rendered a divine reality a matter of belief or hope, rather than direct experience (Griffin 1988).

Weber's analysis has proved enduring and influential, and many contemporary theorists seem to agree that what we now call modernity has an undeniable ability to 'kill the spirit' (Flanagan 2010, 1). However, some other scholars have argued this is becoming an increasingly outdated view, claiming that we are in fact entering a new, more magical, modernity that embraces the irrational, the intangible and the uncanny (see, for example, Bennett 2001, Jenkins 2000, Krieger 1981, Partridge 2002 and 2004, Ruickbie 2006). Such accounts challenge the Weberian description of disenchantment as a unidirectional and universalising tendency in modernity (Jenkins 2000), suggesting that 'outside the iron cage there is another enchanted garden' (Ruickbie 2006, 124). The 'massive spiritual shift' described above, with its growing interest in the more otherworldly, esoteric and occult, reflects this trend (Partridge 2004). There is, suggests Jenkins (2000), evidence therefore of 'decidedly modern' *re*-enchantments which must be recognised as an integral element of modernity.

My ongoing research suggests this new magical modernity is a key part of the contemporary spiritual landscape, and one which should be acknowledged in critical social science analysis if we are to fully understand the impact of spirituality in the lives of those who practice it. As part of that project I have begun to take a more critical look at spirit's touch in our contemporary world and the consequences arising. It is my belief that the spiritual evolution we are witnessing now is premised upon a *re*-enchantment of the world, allowing an active role for spirit to touch the lives of participants in numerous ways. This means the individual practitioner comprehends the world in radically different ways, and the consequence for us as social scientists is that we might have to embrace Einstein's illusion of 'reality' more firmly, and open up our interpretations to acknowledge a place for more fluid, malleable and enchanting possibilities.

Exploring Spiritual-Material Touch

> When we change the way we think about the universe, we change the way
> we act upon the universe. The universe then changes the way it acts upon us.
> (Kaiser 2000, 10)

In order to touch or be touched by spirit it is necessary to establish some tangible way of initiating that contact. A popular means of grasping an elusive and unknowable spirit is through the incorporation of material objects into ritual practices designed to open a channel between the 'material' and 'immaterial' worlds. Here particular objects become offerings to spirit, or become possessed by spirit as a mechanism – often in the form of sacrifice or gifts – to help spirit work in the earthly plane. The centrality of material proxies for immaterial spirit in such settings is significant, yet the objects used in rituals and offerings do not just embody or represent spirit at that time, their effects extend well beyond it (Espirito Santo 2010). Through this 'spiritualisation' of material objects (Eipper 2007) practitioners' relationships with those objects and their meaning in the world change. Touching spirit through the material world therefore has consequences, as the function of objects used in this way can change and even change society itself (Graves-Brown 2000). The result can be very powerful, for once an object is attributed powers of its own it becomes a force, with a capacity to touch others:

> It doesn't require belief in a phenomenon to be affected by it. Christian missionaries destroyed the pagan idols of indigenous peoples around the world not from fear of being converted, but because the *power* of those objects had to be destroyed. (Eipper 2007, 256)

The warning being then, that whilst we do things with objects, 'objects also do things to us' (Buggeln 2009, 357).

Holloway (2003) explores the power of both spiritualised commodities – such as candles and crystals, and much more protean everyday objects, such as televisions and sofas – during the act of enframing 'a space-time' for spiritual insight. For Holloway (2003), carefully crafted spiritual moments are carved out almost despite the immateriality of spirit, and materiality plays a fundamental part in meeting that immateriality. However, is a focus on the *objects* – the touching and placing of the candles, the distractions of televisions, the ritual offerings or quasi-religious props – diverting our gaze from another part of the relationship being enacted? Holloway's objects are organised as a stage for meditation, to receive messages from spirit. Other enframings of such space-times might be the act of holding a crystal to 'tune into' otherwise intangible universal spiritual energies, or the casting of rune stones to receive an otherworldly message. However, a more detailed investigation would question whether this is the full extent of the objects or agents we are dealing with in relation to a contemporary enchanted spirituality (MacKian 2012). Such accounts only focus on the spaces and objects we recognise as part of habitual 'reality' – the candles, crystals and rune stones, and the hands and body of the individual manipulating them. What other objects or agents might we need to incorporate into our analysis to ensure we capture moments of contact in spiritual-material space which might be overlooked by simply observing what people do with material things?

By way of illustration, I now explore two moments of spiritual-material touch. The first example focuses on the experiences of Mark Hankin, who runs his own tarot reading and ghost hunting business in the North West of England. A civil engineer by training, he told me his first profound contact with spirit came when he was in his teens and was visited in hospital by a healing angel in the guise of a staff nurse. In the example I draw on here we see the hand of spirit acting quite literally in the material world, giving him a very tangible feeling of having being touched by spirit. In the second example, I look at one participant's experiences of working with spiritual energies. 'Penny' (a pseudonym) runs a private clinic in the North West of England with her family, offering chiropractic treatment, massage, bush flower therapies, homeopathy and acupuncture. She has always felt the energies of nature, works closely with spirit guides and began developing her spiritually-based healing skills following a diagnosis with cancer in 1995. Her encounters with spiritual energies demonstrate that spirit is also understood to touch in ways which are less immediately visible. Yet specific material consequences following her spiritual-material enactments are interpreted by Penny to be the direct result of, and further evidence of, spirit's ability to touch our world.

Mark: recognising spiritual-material touch

> I opened up my tarot box and on top of my cards was a small white feather. It couldn't have blown in or been put there by anyone. (Mark)

The idea of imbuing particular objects with otherworldly powers, or using ritual objects in establishing a link to spirit is commonplace amongst practitioners. For many spiritual seekers tarot cards are one material proxy that might be used to open a channel and establish tactile contact with spirit. Individual cards may be held and meditated over or placed on a personal altar to invoke particular energies in the person's life. Or the cards may be carefully shuffled and laid out in a spread to bring a message to another person in a reading. The importance of touch in achieving the spiritual connection is reflected in the practice common to many card readers of never allowing others to touch their cards, for fear of the 'wrong' energies being transferred. Alternatively, other readers may insist the sitter holds the cards to impart 'their energies' onto them. Whilst these two practices may seem contradictory, they are nonetheless both based on the same principle that touch – understood as moments of contact (either physical or as a less tangible feeling of communication) – initiates and maintains the relationship and link with spirit. Tarot cards, in essence, are routine agents in the coproduction of spiritual-material spaces between the reader and the sitter. However, the appearance of the white feather in Mark's tarot box suggests we may need to reconceptualise the processes and agencies through which spiritual-material spaces might be coproduced around particular (im)material objects.

For Mark, the gift of a feather in his box of cards did not come from another person, neither was it put there purposively by another human hand. It was not part of *his* construction of material channels or proxies designed to reach out to spirit. It could only be a gift from spirit. In this example, the material was not knowingly placed by earthly forces, but apparently by the immaterial itself. Mark was not the one reaching out using ritual objects – as he might when doing a tarot reading – *spirit* was the active agent here. What was significant for Mark was that spirit alone could have touched his carefully guarded box. Through allowing spirit agency, we see the material feather becoming immaterial, the immaterial spirit becoming material as symbolic of spirit's presence and agency in the earthly world. Evidence that spirit can touch the worlds of believers in material ways.

Mark had repeatedly asked for a feather from spirit as proof that his connection was 'real'. The appearance of feathers is a frequently mentioned everyday example of being 'touched by spirit' or 'knowing spirit is there', because they are interpreted as a physical representation of the presence of angels. Such interpretation requires the blurring of the boundary between apparently distinct conceptual categories – the tangible 'earthly' and the ethereal 'otherworldly'. A practitioner might acknowledge the feather fell from a bird, but will also say it did so to show them spirit was there. This suggests there is in fact no distinction, but instead mutable landscapes of touching and being touched at the interface between the earthly and the otherworldly. It is this very process which makes spirit more tangible, and such moments of contact so profound.

Spirit thereby erupts into the earthly world if individuals recognise it as such, and this requires new conceptualisations of the 'real'. For those attuned to the possibility of contact with spirit, they expect to see, and do see, material signs of this, such as Mark's feather in the tarot box. But whilst materiality might represent temporary touchstones of the intangible immaterial in tactile, knowable ways, as we will see in Penny's example below, there are also other moments of contact which do not rely so directly on the materiality of things. In either case, the meaning for the practitioner of such pathways opening up is that they are literally being 'helped' by spirit (Bennett and Bennett 2000). They are interpreted by those experiencing them as empowering signs that spirit is supporting them and recognising the struggles in their daily lives. In these cases, whether or not these are 'really' spirit objects and manifestations is less important than the *belief* that they are – and the consequences that follow. Because people see these spiritual-material moments as signs of spirit effecting a material change in this world it offers for them the possibility of working *with* spirit to produce further changes in the material world around them, as the following example from Penny demonstrates.

Penny: Working with spiritual-material energies

> If I am out walking on my own I can go into a semi meditative state and get
> my spirit guides to touch me… To walk through beautiful countryside with my
> spirit guides holding my hands fills me with contentment. To hear the chatter
> of flowers, feel the power of the trees, catch a glimpse of something small and
> bright darting through the undergrowth on the edge of my vision reminds me that
> there is a lot more to this world than just what my five senses tell me. (Penny)

The relationship between spirit and the individual once established is an active
one, with the individual being able to work with spirit and channel its 'energies' in
various ways. Whilst practices such as Reiki or hands on spiritual healing might
be the most obvious tactile ways in which this might occur, there are other less
obvious ways in which practitioners experience and utilise spirit's touch.

As the quote from Penny above shows, she had various connections with spirit,
through her personal guides and a range of nature spirits. She used a combination
of obvious material proxies and less visible means of getting in touch with these
spiritual energies in her work as an Australian bush flower essence therapist.
Material props included a pendulum, divination cards, hands-on massage and the
essences themselves. However she also had other ways of channelling spiritual
energies outside the consultations themselves, including a number of strategies to
bring the touch of the otherworldly to bear on her earthly plans:

> If things are a bit slow business wise I will imagine people picking up their
> phone and dialling our number. Within a couple of hours one or two people will
> ring in… If we are doing promotional work and people are walking by instead
> of stopping, drawing on spirit to project a pink glow all around the promotional
> stand can really make people look and take notice. (Penny)

Whilst Penny's tactics may sound rather mercenary and based on personal gain,
she was keen to point out that through these techniques spirit would be directing
people to her who *needed* help. It was a win-win situation: her business would
flourish, and many more individuals would be helped towards the healing they
felt in need of. Penny was able to channel spiritual energy in this way because
she experienced her daily life through numerous overlapping layers of connection
between the otherworldly and the earthly. She had an almost constant awareness
of nature communicating to her, as well as a strong relationship with a number
of spirit guides. Penny's everyday reality was therefore a complex mix of very
earthly phenomena – such as consultations, telephone conversations and walks
in nature; and inherently otherworldly phenomena, including her spirit guides,
nature spirits and the use of spiritual energies to bring about change in her material
circumstances. The way she experienced her world constantly served to reinforce
and demonstrate to her the power of spirit's touch. Although I have used the
example of her business to illustrate this, there were other ways in which she drew

on such energies outside that setting, such as providing readings and healing in a voluntary capacity; spirit's touch was not exclusively linked to the bottom line.

Reflections on Spiritual-Material Encounters

Mark's world was touched by spirit in a literally tangible way, with the appearance of the white feather. For Penny, the power of spiritual energies over her business strategies may appear more subtle and less immediately tangible, but the results for her are just as consequential. Both types of spiritual-material encounter are illustrative of the way in which spirit's touch in our modern world is both felt and seen by those who seek a relationship with it. I now reflect on how these moments of contact and emerging relationships of touch ensure spirit becomes a tangible part of the 'realities' practitioners live within.

Part 1: Feeling Touch

> There are forces that surround us and who are a resource and help in all aspects
> of our lives. (Princess Märtha Louise, quoted in Duffy 2007)

Bingham (2006) argues that the co-production of social life is not simply about relations between people; this co-production is also often between people and 'things'. Although Bingham was writing about bees, butterflies and bacteria, spirit in the form of the feather for example, could quite reasonably represent just another such 'thing'. Just like Penny calling on the energy of spirit around her marketing stall, the spiritual for these practitoners is 'everywhere and immanent in every moment of experience' (Holloway 2003, 1942) and is part of the co-produced ways they sense, experience and interpret the world. To feel this co-production is therefore to acknowledge the persuasive 'reality' of the agency of spirit, and once the individual *feels* this touch it can have profoundly therapeutic and empowering effects.

Yet to experience such encounters and feel that touch requires the telling of a 'dramatically different story' to the one of Western disenchantment which dominates the literature on contemporary spirituality (Johnson and Murton 2007). Contemporary alternative spirituality is often dismissed as ineffectual, because, it is claimed, it tackles the symptoms of society's malaise rather than the underlying causes (see, for example, Bruce 2002, Carrette and King 2005). The argument is that it facilitates temporary, individualised moments of euphoric happiness whilst leaving the essentially isolating weakness of modern society unchallenged. However, for those reporting a connection with spirit through their spiritual practices, the profound impact of such experiences offers a new possibility for reading the world in a different way. As a result of really 'feeling' a connection with spirit, individuals find alternative relationships with the world around them which becomes empowering for them, but is also perceived as having potential ramifications for the wider social condition.

> It has to start with the individual... Now on a bigger scale instead of one person lets take 100,000 people who are like minded in the way I love life, I love people, I am safe, I am protected and all of these people send out positive thought... not one bad thought, not one bad word. Now let's take that one step further the UK has what 55 million people on it, what if every one of those sent out the same thoughts, within 80 to 100 years you would have a UK with no anger, greed, war, jealousy. (Kevin)

Or put in academic terms:

> the concept of personal responsibility can also include responsibility to others, and thus, has social consequences. Inner awareness may produce awareness of the interconnections of personal conditions with larger social conditions... enlightened self-interest can transform into concern for others... under some conditions, 'transformation of self' may be a catalyst for political action. (Finley 1991, 35)

Exploring the potential impact of this is beyond the scope of this chapter (see MacKian 2012 for more exploration of the 'infrapolitics' of contemporary spiritual experience and practice); suffice to say, that the material impacts of feeling spirit's touch demand closer attention than hitherto applied.

Part 2: Seeing Touch

> The real voyage of discovery consists not in seeking new landscapes but in having new eyes. (Proust 2000, 291)

Each small-scale example of individuals or groups effecting material change in the world around them as a result of their relationship with spirit begins to map new topographies onto the landscapes we thought we knew. New topographies which are unlikely to overthrow the disenchanted world we believe we live, but certainly contribute towards an unsettling of the firm foundations of the material world we thought we knew. Lee (2001, 155) warns that contemporary notions of spirituality must not be romanticised as 'ideal', or 'accepted unquestionably as the way forward in the twenty-first century'. Neither should they, however, be simplistically dismissed – as for example Carrette and King (2005) do – as failing to provide any answers at all. These practitioners, and many millions like them across the Western world (Ivakhiv 1996), feel and see the touch of spirit in myriad ways around them, and therefore sense the world in a way social science currently struggles to engage with adequately.

Seeing the impact of encounters with the otherworldly is part of the everyday landscape for these spiritual practitioners, ensuring the continued material and sensed presence of the otherwise intangible. For these practitioners spirit therefore touches the modern world in many more ways than the disenchantment thesis acknowledges, not only through specific spiritual sites and buildings, but more broadly across 'spaces for work, play, learning, socialising, intimacy, movement and rest' (Ivakhiv 2001, 232). There is no doubt that the illusion of what might be considered to be 'reality' shifts as a result of seeing these spiritual-material encounters, and this has important implications for how we imagine the world and the way that world works (Dewsbury and Cloke 2009). We need therefore to suspend judgement over the veracity of claims made by these spiritually inspired individuals, and to reflect instead critically upon the impact of seeing their particular reality on their lives and the lives of others around them. It is therefore time we started to look at these experiences of spiritual touch with new eyes.

Concluding Thoughts on the Spiritual-Material Spaces of Touch

> However other-worldly the esoteric and the occult may seem, they gain their credibility from their relationship with the real world, whether through their ability to manipulate it or to live better within it or simply to understand it. (Chris and Bartolini 2010, 14)

Spirit's touch – whether we believe in it or not – is embedded in our physical, social and personal worlds. It was undeniably 'real' in the perception and phenomenal world of my research participants, it had social significance and altered their behaviour in this world. Without contact from spirit, Mark would not have established his Tarot business; and without spirit's guidance, Penny would not have trained as an essence therapist. Yet much of the academic commentary around contemporary alternative spirituality has been produced with very little space given to how such spiritual encounters manipulate the contours of our material worlds. In this chapter I have suggested that by seeing and feeling spirit's touch in their everyday worlds practitioners encounter the emergence of new landscapes of experience, 'heterotopic' sacred spaces, constructed and touched by many actors, including the other-than-human (Ivakhiv 2001).

Whilst Mark's experience with the feather and Penny's spiritually enhanced marketing techniques may appear to the uninitiated to be simply *too* otherworldly to countenance, they are far from unique and certainly not alone. Indeed, Partridge (2004) suggests there has been a general 'occulturation' of Western popular culture, reflecting a more widespread interest in the esoteric, the occult and the otherworldly. Magic and spirits have been brought into the mainstream through the mass media (Chris and Bartolini 2010), leading to the supernatural – in the form of Buffy the Vampire Slayer or Percy Jackson the demi-god – to be seen as 'cool'. But enchantment is not simply about being enchanted by mystical stories or mythical creatures; it can also have very real impacts on the personal morals, values and actions of those it touches (Chris and Bartolini 2010). As Chris and Bartolini say, the otherworldly has a very tangible relationship with our world, because it can, and does, manipulate it, and can help people to live in it and understand it better. I suggest it is time then to re-evaluate the way in which spirit touches our everyday world.

> I would now say I have as much faith in the spirit world and the source as I do in any living human, if not more, because spirit have never let me down. (Kevin)

As Kevin's quote suggests, the touch of spirit in everyday life is something that can cultivate a sense of connection in a way that our individualised, demarcated and increasingly privatised lives often appear to deny us. There is 'a widespread experience of insecurity in late modern society' (de Groot 2006, 97), and we must therefore take seriously any opportunity for satisfying the desire for ontological security (Bauman 2000). An ontologically secure person 'will encounter all the hazards of life, social, ethical, spiritual, biological, from a centrally firm sense of his own and other people's reality and identity' (Laing 1969, 39). The spiritual stories told by my research participants suggest strongly that these individuals believe themselves to have reached such a position of security directly as a result of contact with spirit and of feeling and seeing evidence of that in their world. Highlighting the importance of touch allows us to see how these relations are solidified in the materiality of everyday realities.

It should be acknowledged, however, that I am focusing here on the positive effects of being touched by spirit; and I recognise that such encounters can also manifest in more disturbing or frightening ways. Nonetheless, my participants focused on the positive, liberating and empowering aspects, and that is why the possibly darker side to otherworldly encounters is absent from this particular analysis. Of course some experiences – such as visits from angels – may be the result of psychotic episodes or other physical, neurological or mental conditions (see, for example, d'Orsi and Tinuper 2006, Bisulli et al 2004), and might under clinical scrutiny be interpreted as having pathological origins. However, it has been suggested that even events which might appear to be psychotic episodes may in fact occasionally be manifestations of non-pathological spiritual experiences. The relationship between spiritual experiences and psychosis is therefore a complex and currently poorly understood one, and not one I have room to explore here. From a social science perspective (rather than a psychiatric or clinical one), I am willing to accept these experiences of being touched by spirit as inherently positive *for these particular individuals*.

I am calling therefore, for a re-enchantment of the spiritual discourse which allows us to explore the way in which enchanted moments of contact with spirit manifest in everyday encounters within and through everyday practices, and depicts something of the touching enchantment of magical modernity. In this chapter I have therefore argued for the consideration of touch in our understandings of spiritual relationships and practice, for it is only when we acknowledge how profoundly spirit touches people's lives that we understand from a social science perspective the full impact it has on those lives. For the modern world is still touched by spirit and layered with the magical intangibility of other worlds; but social science needs to look with new eyes to see this, and to understand the sociological implications of the presence of such enchanted realities.

References

Bauman, Z. 2000. *Liquid Modernity*. Cambridge: Polity Press.

Beck, U. 1994. The reinvention of politics: towards a theory of reflexive modernization, in *Reflexive modernization*, edited by U. Beck, A. Giddens and S. Lash. Cambridge: Polity Press, 1–55.

Beck, U. and Beck-Gernsheim, E. 2002. *Individualization: Institutionalized Individualism and Its Social and Political Consequences*. London: Sage.

Bennett, J. 2001. *The Enchantment of Modern Life: Attachments, Crossings, and Ethics*. Oxford: Princeton University Press.

Bennett, G. and Bennett, KM. 2000. The presence of the dead: an empirical study. *Mortality*, 5(2), 139–57.

Berger, P.L. 1970. *A Rumour of Angels: Modern Society and the Rediscovery of the Supernatural*. London: Anchor Books

Bingham, N. 2006. Bees, butterflies, and bacteria: biotechnology and the politics o nonhuman friendship. *Environment and Planning A*, 38, 483–98.

Bisulli, F., Tinuper, P., Avoni, P., et al. 2004. Idiopathic partial epilepsy with auditory features (IPEAF): a clinical and genetic study of 53 sporadic cases. *Brain*, 127, 1343–52.

Bruce, S. 2002. *God is Dead: Secularization in the West*. Oxford: Blackwell.

Buggeln, G. 2009. A word on behalf of the object. *Material Religion*, 5(3), 357–58.

Carrette, J. and King, R. 2005. *Selling Spirituality: The Silent Takeover of Religion*. London: Routledge.

Chris, R. and Bartolini, N. 2010. *Esoteric Economies: Final Report*. Milton Keynes: The Open University.

Coakley, A.B. and Duffy, M.E. 2010. The effect of therapeutic touch on postoperative patients. *Journal of Holistic Nursing*, 28(3), 193–200.

de Groot, C.N. 2006. The church in liquid modernity: a sociological and theological exploration of a liquid church. *International Journal for the Study of the Christian Church*, 6, 91–103.

Dewsbury, J.D. and Cloke, P. 2009. Spiritual landscapes: existence, performance and immanence. *Social and Cultural Geography*, 10(6), 695–711.

D'Orsi, G. and Tinuper, P. 2006. 'I hear voices…': from semiology, a historical review, and a new hypothesis on the presumed epilepsy of Joan of Arc. *Epilepsy and Behavior*, 9, 152–57.

Duffy, M. 2007. Princess Martha tells us why she has set up business with Angels. *Digital Journal*, <www.digitaljournal.com/article/210452>.

Eipper, C. 2007. Moving statues and moving images: religious artefacts and the spiritualisation of materiality. *Australian Journal of Anthropology,* 18(3), 253–63.

Espirito Santo, D. 2010. Spiritist boundary-work and the morality of materiality in Afro-Cuban religion. *Journal of Material Culture,* 15(1), 64–82.

Finley, N.J. 1991. Political activism and feminist spirituality. *Sociological Analysis*, 52(4), 349–62.

Flanagan, K. 2010. Introduction, in *A Sociology of Spirituality,* edited by K. Flanagan and P.C. Jupp. Aldershot, England: Ashgate, 1–21.

Foltz, T.G. 2000. Women's spirituality research: doing feminism. *Sociology of Religion*, 61(4), 409–18.

Giddens, A. 1994. Living in pot-traditional society, in *Reflexive Modernization*, edited by U. Beck, A. Giddens and S. Lash. Cambridge: Polity Press, 56–109.

Graves-Brown, P.M. 2000. Introduction, in *Matter, Materiality and Modern Culture*, edited by P.M. Graves-Brown. London: Routledge, 1–9.

Griffin, D.R. 1988. *Spirituality and Society: Postmodern Visions*. Albany, NY: State University of New York Press.

Hart Wright, S. 2002. *When Spirits Come Calling: The Open-minded Sceptic's Guide to After-death Contacts*. Nevada City, CA: Blue Dolphin Publishing.

Hay, D. and Nye, R. 2006. *The Spirit of the Child*. London: Jessica Kingsley.

Heelas, P. 2008. *Spiritualities of Life: New Age Romanticism and Consumptive Capitalism*. Oxford: Blackwell.

Heelas, P. and Woodhead, L. 2005. *The Spiritual Revolution: Why Religion is Giving Way to Spirituality*. Oxford: Blackwell.

Hetherington, K. 2003. Spatial textures: place, touch, and praesentia. *Environment and Planning A*, 35, 1933–44.

Holloway, J. 2003. Make-believe: spiritual practice, embodiment, and sacred space. *Environment and Planning A*, 35, 1961–74.

Hoyez, A.C. 2007. The 'world of yoga': the production and reproduction of therapeutic landscapes. *Social Science and Medicine*, 65, 112–24.

Ipsos Reid. 2006. *Canada Speaks Survey*. Sympatico/MSN. <www.marketwire.com/press-release/Ipsos-Reid-618230.html>.

Ivakhiv, A. 1996. The resurgence of magical religion as a response to the crisis of modernity: a postmodern depth psychological perspective, in *Magical Religion and Modern Witchcraft*, edited by J.R. Lewis. Albany, NY: New York University Press, 237–67.

Ivakhiv, A.J. 2001. *Claiming Sacred Ground: pilgrims and politics at Glastonbury and Sedona.* Bloomington, IN: Indiana University Press.

Jenkins, R. 2000. Disenchantment, enchantment and re-enchantment: Max Weber at the Millennium. *Max Weber Studies*, 1, 11–32.

Johnson, J.T. and Murton, B. 2007. Re/placing native science: indigenous voices in contemporary constructions of nature. *Geographical Research*, 45, 121–29.

Kaiser, L.R. 2000. Spirituality and the Physician Executive. *The Physician Executive*, March-April, 6–13.

Knibbe, K. and Versteeg, P. 2008. Assessing phenomenology in anthropology: lessons from the study of religion and experience. *Critique of Anthropology*, 28(1), 47–62.

Krieger, D. 1981. *Foundations for Holistic Health Nursing Practices*. Philadelphia: JB Lippincott Company.

Laing, R.D. 1969. *The Divided Self*. London: Penguin Books.

Lee, H. 2001. Towards a critical politics of spirituality *Critical Psychology*, 1, 153–157.

Levin, J. 2003. Spiritual determinants of health and healing: an epidemiological perspective on salutogenic mechanisms. *Alternative Therapies in Health and Medicine*, 9(6), 48–57.

MacKian, S. 2010. In possession of my senses? Reflections from social science on engaging with the otherworldly. *Parananthropology: Journal of Anthropological Approaches to the Paranormal*, 2(1), 27–29.

MacKian, S. 2012. *Everyday Spirituality: Social and Spatial Worlds of Enchantment*. London: Palgrave Macmillan.

Marta, I.E.R., Baldan, S.S., Berton, A.F., Pavam, M. and da Silva, M.J.P. 2010. The effectiveness of Therapeutic Touch on pain, depression and sleep in patients with chronic pain: clinical trial. *Revista Da Escola De Enfermagem Da USP*, 44(4), 1094–1100.

Miller, W.R. and Thoresen, C.E. 2003. Spirituality, religion, and health: an emerging research field. *American Psychologist*, 58(1), 24–35.

Monzillo, E. and Gronowicz, G. 2011 New insights on therapeutic touch: a discussion of experimental methodology and design that resulted in significant effects on normal human cells and osteosarcoma. *Explore: The Journal of Science and Healing*, 7(1), 44–51.

Moore, D.W. 2005. *Three in Four Americans Believe in Paranormal: Little Change from Similar Results in 2001*, Gallup.

Norris, P. and Inglehart, R. 2004. *Sacred and Secular: Religion and Politics Worldwide*. Cambridge: Cambridge University Press.

Partridge, C. 2002. The disenchantment and re-enchantment of the West: the religio-cultural context of contemporary Western Christianity. *The Evangelical Quarterly*, 74(3), 235–56.

Partridge, C. 2004. *The Re-enchantment of the West, Volume 1: Alternative Spiritualities, Sacralization, Popular Culture, and Occulture*. London: T&T Clark.

Paterson, M. 2009a. Introduction: Re-mediating touch. *Senses and Society*, 4(2), 129–40.

Paterson, M. 2009b. Haptic geographies: ethnography, haptic knowledges and sensuous dispositions. *Progress in Human Geography*, 33(6), 766–88.

Proust, M. 2000. *In Search of Lost Time Volume 5: The Captive and the Fugitive*, Translated by C.K. Scott Moncreiff and T. Kilmartin. London: Vintage Books.

Roof, W.C. 1999. *Spiritual Marketplace: Baby Boomers and the Remaking of American Religion*. Princeton, NJ: Princeton University Press.

Ruickbie, L. 2006. Weber and the witches: sociological theory and modern witchcraft. *Journal of Alternative Spiritualities and New Age Studies*, 2, 116–30.

Schneider, M.A. 1993. *Culture and Enchantment*. Chicago: University of Chicago Press.

Spencer, N. and Alexander, D. 2009. *Rescuing Darwin: God and Evolution in Britain Today*. London: Theos.

Thrift, N. 2004. Movement-space: the changing domain of thinking resulting from the development of new kinds of spatial awareness. *Economy and Society*, 334, 582–604.

Tyler, L.C. 2009. *A Very Persistent Illusion*. London: Pan Books.

Vellenga, S.J. 2008. Hope for healing: The mobilization of interest in three types of religious healing in the Netherlands since 1850. *Social Compass*, 55(3), 330–50.

Vickers, C.R. 2008. Healing touch and therapeutic touch in the psychiatric setting: implications for the advanced practice nurse. *Visions: The Journal of Rogerian Nursing Science*, 15(1), 46–52.

Voas, D. and Bruce, S. 2007. The spiritual revolution. Another false dawn for the sacred, in *A Sociology of Spirituality*, edited by K. Flanagan and P. Jupp. Aldershot, England: Ashgate, 43–61.

Wirzba, N. 2003. *The Paradise of God: Renewing Religion in an Ecological Age*. New York: Oxford University Press.

Wylie, J.W. 2009. Landscape, absence and the geographies of love. *Transactions of the Institute of British Geographers*, 34(3), 275–89.

Index

Milton Keynes UK
Ingram Content Group UK Ltd.
UKHW031145141024
449569UK00024B/1051